MILK CHURNS TO MERRY-GO-ROUND

A century of train operation

R. T. MUNNS

DAVID & CHARLES
Newton Abbot London North Pomfret (VT)

British Library Cataloguing in Publication Data

Munns, R. T.
Milk churns to merry-go-round: a century of train operation.
1. Railways — Great Britain — Freight — History
I. Title
385'.24'0941 HE3018

ISBN 0–7153–8666–2

Typeset by Photographics Ltd
and printed in Great Britain
by Redwood Burn Limited, Trowbridge, Wilts
for David & Charles Publishers plc
Brunel House Newton Abbot Devon

Published in the United States of America
by David & Charles Inc
North Pomfret Vermont 05053 USA

CONTENTS

TO IRIS

"The life of every man is a diary in which he means to write one story and writes another: and his humblest hour is when he compares the volume as it is with what he hoped to make it"

(James M. Barrie)

FOREWORD

by Sir Henry Johnson, K.B.E.
Chairman of British Rail 1968–1971

Some 60 years ago, two young men were on night duty in the Down Decoy marshalling yard at Doncaster – Reginald Munns and Henry Johnson. This was the beginning of a long and, for much of the time, close and happy association, particularly in the days of the LNER and later the Eastern Region of British Rail.

Early training and experience at stations and depots laid the foundation of Munns' success. It was a tough school in which his dedication and hard work soon brought him to the notice of those in authority.

In the Chapter 'Railways in Wartime' Reginald Munns paints a vivid picture of the burdens on the railways. My association with him during those years was often close. He was a tower of strength, never flinching and seldom at a loss in dealing with problems, many almost insuperable. Perhaps above all his staff never lacked decision, firmly given and always with his full backing and support.

Any impression that Reginald Munns was in any way hidebound or limited in outlook would be entirely false. He was an innovator – Merry Go Round the perfect example – seldom short of ideas, vigorous and unremitting in his pursuit of them. Woe betide any member of his staff who was found lacking in Reginald Munns' pursuit of excellence.

His reputation as a freight operator was firmly established early in his career. Later he showed a marked aptitude for negotiations with major customers whose representatives were quick to appreciate his grasp of the essentials and clear exposition of his case. The acceptance of new, even revolutionary methods, owes much to his powers of persuasion.

Reginald Munns was born and bred in the tough school of

GN discipline, father and grandfather before him. Not for him the 'purple and fine linen' of the softer establishments. Later he was successful on the LNER and Eastern Region of British Rail against tough competition. He earned and deserved the respect and liking of colleague and competitor alike, trusted completely by all.

I agreed gladly when he asked me to write this foreword. Sadly, before it could be completed Reginald Munns died and it is at the wish of Iris, his widow, that I complete the task. If in some measure it can rank as a memorial, it could not honour a better man.

INTRODUCTION

In a full working life on the railway much of my work was in railway activities, particularly freight, and train and traffic control, not often noticed in railway books, and indeed there appeared to be a distinct gap which this book attempts to fill. Thus the following pages record broadly but often in detail (and sometimes autobiographically) the development of railway operations over a century.

It is based largely on the practices as I saw them on the Great Northern and the Great Central and partially on the Great Eastern, with a purposeful emphasis on the movement of freight which from 1825 brought greater industrial prosperity to this country and, indeed, throughout the world. Historically, the 'carriage of goods, coal and merchandise' as it was described in years past, provided two-thirds of the railway revenue – the bread-and-butter – and thus, to a certain extent, subsidised the passenger. Now on modern BR the situation is reversed and the passenger has to pay his way.

My own early years on the railway were thankfully disciplined at a time when 'you did as were told', to report for duty at this or that station without question, and to pay respect to superiors not forgetting the use of 'Sir'! Consequently, before my first promotion I had been through the Great Northern's Signal & Telegraph School at Retford and which was known as 'The Rooms' (where the instructor used a baton on the legs of those youngsters he thought deserved it!) and, after a six-weeks course was posted to the Yardmaster's Office at Colwick, and thereafter to Grantham, King's Cross, Peterborough New England (four posts), Boston, Marylebone, (for the Wembley Exhibition), Nottingham, Doncaster (three posts), and Bradford. Thereafter I was in London and the provinces as Assistant Controller, Section Controller, Assistant Yardmaster, Stationmaster/Yardmaster, Deputy Chief Controller at District and at Headquarters level, Chief Controller and Head of Passenger & Freight, Head of Freight at Headquarters and, finally, Freight Assistant, Eastern Region General Manager's Office. There I had the responsibility for the planning, operation and monitoring of freight services originating in, terminating in or passing through the Region, and which involved close co-operation not only with other railway Regions but also with the National Coal Board, the Central Electricity Generating Board, the then British Iron & Steel Federation, and the British Sugar Corporation.

Much of the information and data has been collated over the years and I gratefully acknowledge ready assistance and co-operation received from the *Lincolnshire Echo*, British Rail, *Railnews*, the National Coal Board, and the Central Electricity Generating Board at Peterborough, Birmingham and London. Finally my special thanks are due to Arthur Cutts and Charlie Smith.

R. T. MUNNS
Tunbridge Wells

1
THE COUNTRY SCENE

No-one needs to be reminded that, though it was not the first rail way, the Stockton & Darlington Railway was the first public railway to use steam locomotives from the beginning for the conveyance of goods, and the date of 1825 is stamped on many minds. But what is important and relevant is that it created throughout the country, and abroad, a system of linked depots better known as country stations which were associated with adjacent coal mines, ironstone mines, other industries, and agriculture. Indeed, the coming of the railway provided a rural revolution.

The traffic from farmlands proliferated and new markets were opened up. Individual consignments increased to the point where they needed to be gathered into local sidings and from there, as time passed, into larger concentrations or marshalling yards for processing into full train loads for more distant parts. By and large the country station was the prime source of all traffic to urban areas and it received back more of the commodities generated by commercialism. Hitherto the only wheeled vehicles available for the conveyance of merchandise had been the farm cart and the carrier's cart but they became the feeders to the new railway wagon and soon *everything* was carried by the railway, which made the country station a very busy place. The stationmaster was a busy man with considerable responsibility, always ready to use commercial and operating initiative for in the early days there was no train and traffic control office to give direction or carry the can if things went wrong as in later years. There were many such stations but a description of one best known to the author in the early years of this century would surely demonstrate that the working at a country station like Scalford in Leicestershire, on the Great Northern and London & North Western Joint Railway, bore no relation to the antics portrayed in the

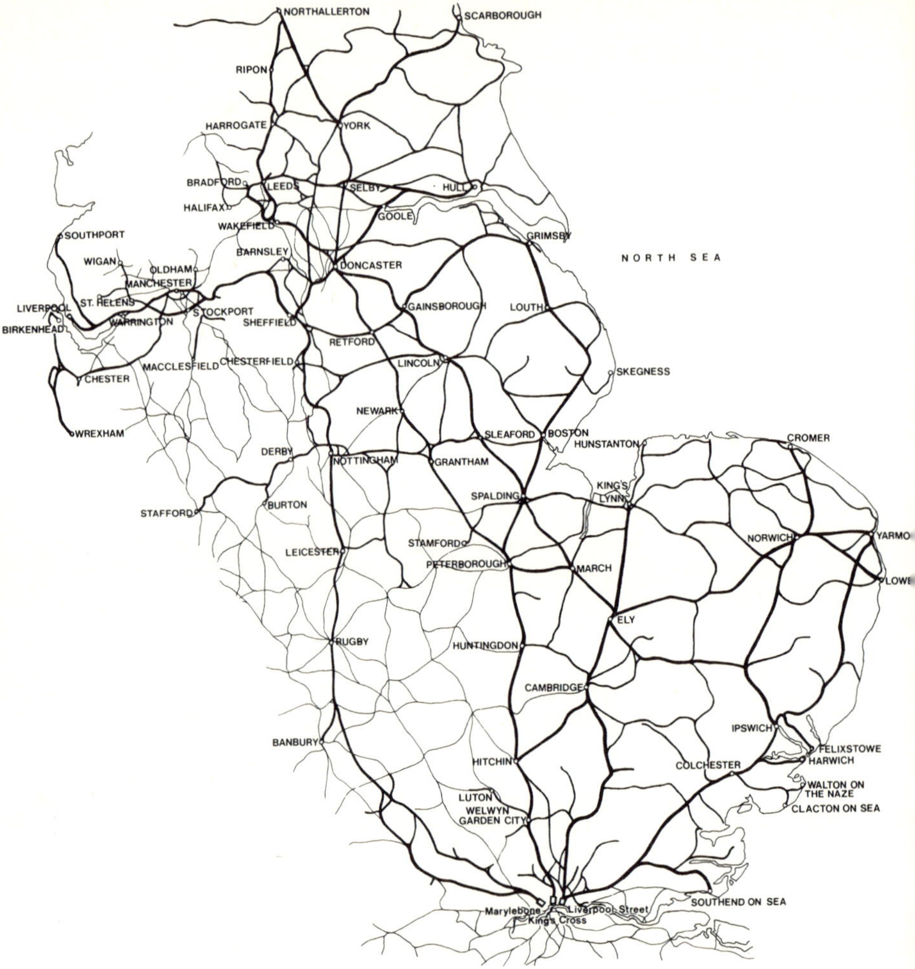

The LNER Southern area and it connections, covering the former GE, GN and GC routes.

classic comedy films *Oh! Mr Porter* and *The Titfield Thunderbolt!*.

The GN & LNW Joint Line was an amalgam of interests resulting in the construction in 1879 of a line between Welham Junction just north of Market Harborough, Bottesford, west of Grantham, and Saxondale, east of Nottingham. From this line the Great Northern constructed a branch in 1882 from

John O'Gaunt (Marefield Junction) into Leicester (Belgrave Road). The commercial interest behind this enterprise was a desire to exploit the wool trade to Yorkshire's woollen mills, the hides and leather for boot making, and the iron ore known to be available in the Leicestershire and Lincolnshire Wolds for movement to the blast furnaces of Holwell, Stanton and Staveley. Passenger traffic was a secondary but nevertheless important consideration. This was the background confronting the author's grandfather in 1882 when he was appointed as the first stationmaster at Scalford. A native of Deal in Kent he had before that appointment been a stationmaster on the Great Northern's East Coast main line. He was succeeded by his son in 1904 who had previously had experience in London, the Midlands and the West Riding. So the two were in charge of the same station for 37 successive years and, in all, 52 years on the Joint Line as stationmasters. If the handbook or rule-book *General Instructions and Regulations for the Executive Department* issued by the Great Northern Railway in 1856 and to which they had to subscribe and adhere is anything to go by, a high standard of character, loyalty, integrity, sobriety, etc, was demanded. As the National Railway Museum at York has it 'Stationmasters held positions of some respect in the communities they served'.

The two stationmasters, in their turn, were activists in more senses than one; they had become a local institution in a period of English, paraffin-lamp-lighted country life, when the coming of railways brought someone who was 'a cut above the average' even if the average was low, and with a wider experience and background furnishing ability to supplement the vocational and extra-mural interests of both the vicar and the schoolmaster. Both were churchwardens and Sunday School superintendents covering 38 years. There were the parish council, church council, Band of Hope, childrens' gala, horticultural show, village institute, the brass (later Silver) band, whist drives, concerts and the like, making up the social life of the village, all of which laid claim for leadership, support or participation, and all on top of the stationmaster's 13-hour day, six days a week. Scalford, like most country stations served a radius of several miles and included the residential and fox-hunting gentry (in the Belvoir country),

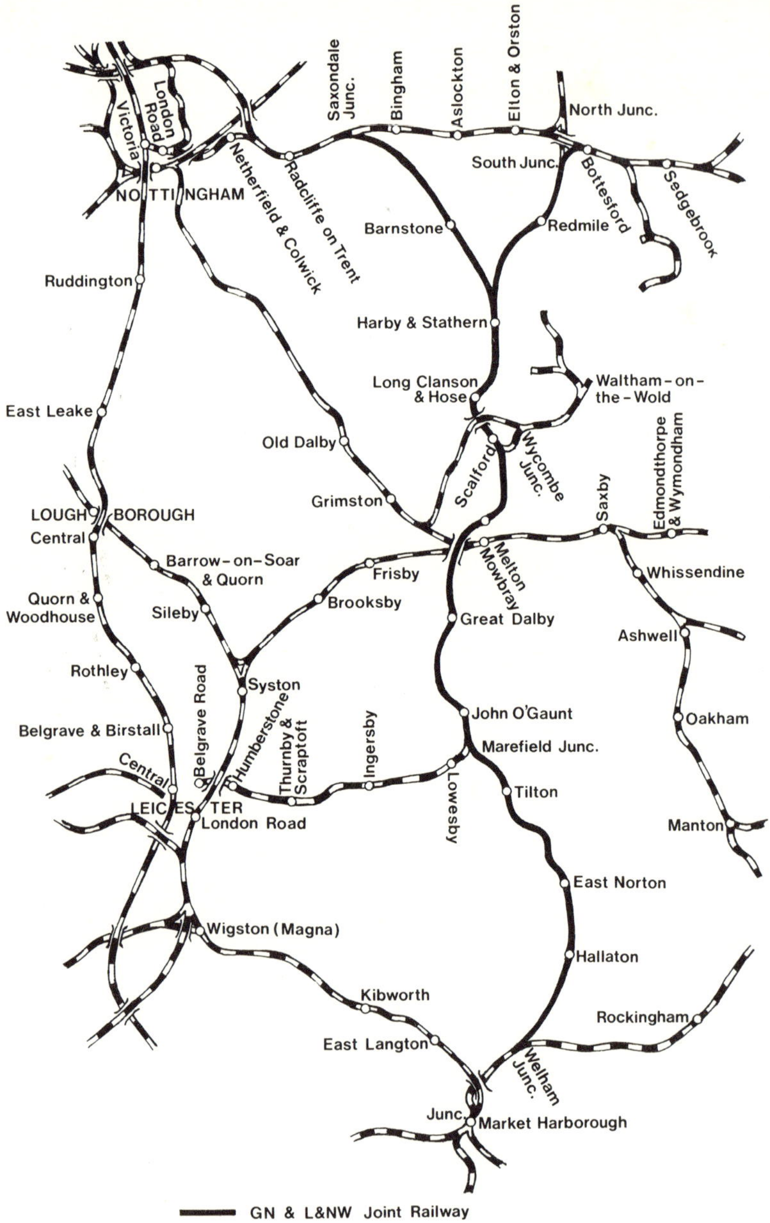

Map of the Great Northern & London & North Western Joint line and its connections.

professional and business people, farmers, and those small acreage farmers categorised in the local directory as 'graziers', as well as the wandering cattle and horse dealers.

Other local inhabitants worked on the non-mechanised farms as waggoners (looking after horses), cowmen, hedgers or just labourers, and others worked in the local brickyard or in the ironstone quarries of Holwell, Eastwell, and Eaton, walking miles across fields or along the many bridle footpaths to and from work. Mobility otherwise was restricted to neighbouring villages on foot or bicycle, the latter guided by the beam of an oil lamp or, if influenced by one-upmanship the much more powerful carbide lamp. The new railway service provided passengers with a wider scope for travel; although not the primary consideration for building the line it did, however, give a service which was better and more frequent than that enjoyed by many a town main line station like Huntingdon, for instance.

The opening up of the Notts/Derbyshire coalfield in 1875, and the resultant group of sorting sidings at Colwick, four miles east of Nottingham, in 1878/79, being enlarged 10 years later, was the foundation of a great and sustained movement of coal in particular, and goods by rail. In those days railway locomotives and ships were fed on coal as were local authority electric power stations and gasworks, not forgetting the domestic consumers when every house had one or more coal fires. Many train loads of coal – not directly revenue earning – had to be moved each week to railway locomotive depots in other parts of the country, and still many more went to the ports for shipment to our maritime stations like Gibraltar, Malta, Aden, and others for fuelling the Royal and Merchant Navy fleets. Consequently the volume of coal moved by rail was tremendous which meant that the sorting sidings at Colwick had to be increased in size and capacity (1889) to become what was reputed as being the largest marshalling yard in the country, though Colwick was never modernised as Whitemoor, Wath and Toton were in later years.

Over the Joint Line, in addition to the through trains from Colwick to Willesden there were others to and from Market Harborough (Welham Yard), Rugby and Northampton, and the goods trains between Colwick/Doncaster/Newark to

Northampton and Leicester (Belgrave Road) as well as the local stations clearing the ironstone mined in the area and which grew to about 624,000 tons a year. For a long time in the latter years of the steam locomotive there were train loads of steam coal passing quite regularly from Rossington Colliery south of Doncaster to Camden LMS locomotive depot.

Although this account is freight orientated it would not be out of place to complete the scene at this typical country station where the passenger services consisted of five or six Great Northern teak-coloured trains each way between Grantham and Leicester, (Belgrave Road) and a similar number of London & North Western plum and spilt milk-coloured trains plying between Northampton (Castle) and Nottingham (London Road, Low Level); the 8.50am from Nottingham conveyed a 'Through coach Melton Mowbray to Euston' arriving at 12.31pm and returning from Euston on the 4.10pm to connect with the 5.50pm from Northampton. The L&NW passenger guards wore a dark blue uniform frock coat trimmed with white piping and they had a leather shoulder pouch for a watch. The first train of the day was at 7.15am and the last at 8.20pm and the stationmaster generally was in attendance all the time – a working day of over thirteen hours, six days a week with no overtime pay or extra remuneration for 'unsocial hours' as would be the case amongst most employees today. In all there were some fifty passenger and freight trains daily in each dirction. The first and last passenger trains of the day to Grantham conveyed the milk for London (Seven Sisters, later re-named Finsbury Park) brought in by farmers and which amounted to about 100 churns daily. Transferring loaded milk churns from farmers' floats to four-wheel barrows, and into the vans on each train was a co-operative effort on the part of farmers, porters and train guards. In the summer season expresses ran between Leicester (Belgrave Road) and Skegness, Mablethorpe, Bridlington, and Scarborough, stopping specially here and there by prior arrangement to pick up or set down parties to and from the seaside.

Scalford station was quite a busy place with coal for local consumption, there being no gas or electricity as alternatives, barrelled beer from the breweries, farm feeding stuffs and produce, grain, hay, straw, sugar beet, round timber, (ie tree trunks), Stilton cheese (of which there was a regular supply to the House of Commons) and not forgetting the truck loads of cattle from Ireland which came to be fattened and later went away on the Wednesday morning special cattle train to Nottingham for slaughter. Amongst the parcels traffic dealt with were large packages, light in weight, which contained the kind of netting used for ladies' veils. These came from Nottingham and were a cottage industry; the consignee-agent distributed contents to clients for making up.

Scalford was a junction station with a single line/branch (built in 1883) 3½ miles long to Waltham-on-the-Wold with gradients of 1 in 50 and 1 in 80 for about a mile to Wycombe Junction where the Midland branch (1887) from Holwell converged into it. From Waltham another three-mile branch continued to the Eastwell and Eaton (1884) ironstone mines and although the latter branches were constructed to get at the ironstone, agricultural traffic to and from the local farms was also dealt with. These two mines produced some 400,000 tons a year.

The great days at Scalford were when the Belvoir pack of foxhounds met in the vicinity, often in the station forecourt, when the hunt specials from London or Nottingham would arrive with passengers and horse-boxes; if too many horse-boxes had to be accommodated not all could get into the carriage dock so that the horses had to be unloaded on the passenger platform. It was all hunting pink then for the gentlemen, with ladies in black side-habits wearing tall hats and veils. They would commandeer all the four-wheeled milk barrows to use as mounting steps and soon the undulating countryside would be ringing with 'tally-ho'! In the late afternoon the mud-splashed hard riders would be making their way back slowly to the station via the bridle-paths across fields to load up for home – to their town houses if they were not maintaining a hunting-box and participating in the Melton hunting social whirl.

Until the first world war an annual horse fair was held at Waltham and there were also steeplechase meetings at nearby Croxton Park. Both attracted a good following with a special service of passenger and horse-box trains laid on between Scalford and Waltham. On such occasions extra staff to deal with horses at Waltham and a ticket inspector and a telegraphist were drafted in to Scalford. There were occasions when batches of German prisoners-of-war were marched from Croxton Park to Scalford to entrain for 'who knows where'.

This was the day of the half-penny postage and many advice notes were posted by the station staff at the station post-box addressed to local farmers and others advising them that a wagon or consignment had arrived and awaited collection and that delay would incur a demurrage payment. In the dark hours of a wintry morning, or the freshness of an early dawn, these advice notes would be in the bag of the one-eyed postman who wore the then uniform shako, trousers tied with string at the ankles. He tramped round the outlying farms, and then, after a smoke of thick twist in the station general waiting room, caught the 7.41am train from Scalford back to Melton Mowbray. Farmers were not then on the telephone and the early morning advice enabled them to plan their day's work and so avoid unnecessary demurrage charges. The signalbox on the platform at Scalford had telegraphic contact with Grantham and Leicester and intermediate stations; one was able to despatch a postal telegram or such could be received addressed to the stationmaster or local resident.

The station buildings at Scalford, like the others on the Joint Line, were far superior to many stations built earlier by the parent companies, being substantially brick-built with canopies over each up and down platforms and with 'General' and 'Ladies' waiting rooms on each side; on at least one occasion the Ladies' waiting room was the venue for a wedding reception! The Ladies' rooms were floored with thick linoleum and furnished with a mahogany table and two long mahogany seats well upholstered with horsehair and the lavatories for both sexes were water-flushed – the 'mod-cons' of the late 1870s and a convenience not vouchsafed to the houses provided for the station staff. The stationmaster's house on the opposite side of the forecourt was detached and standing in its own garden but the architect must have had strange ideas as only one window faced the front and that was the one in the walk-in pantry; even that was covered with gauze! The stationmaster's house at Barnstone, and also Bingham Road, were much superior in design and situation.

Above Leicester Belgrave Road GN station, served by trains off the GN&LNW Joint line. (Eastern Region, BR); *Below* The departure end of Leicester Belgrave Road. (L&GRP/David & Charles)

Above Part of Colwick marshalling yard. (L&GRP/David & Charles); *Below* Peterborough Crescent Junction, just on the London side of Peterborough North station. In the middle distance are Nene sidings while the Midland connection to Peterborough GE station goes off to the right. (L&GRP/David & Charles)

The country station had of necessity to be supervised by a stationmaster who was, so far as railways were concerned, something of a jack-of-all-trades and a master of most. Not only had he to be proficient in signalling and safety rules and regulations by examination but he had to have a wide knowledge of how various traffics had to be dealt with and to instruct staff accordingly. And his accounts had to satisfy the auditors which, in the case of the GN and L&NW Joint Line, had to be done in two ways . . . to Great Northern (LNE) requirements leading up to the month-end balance sheet, and to London & North Western (LMS) requirements leading up to the month-end 'Account Current'. Understandably the relief staff provided by the parent companies to cover holidays and illness did not relish being at a Joint Line station at such times.

The foundation of the railway was the carriage of goods and coal to and from country depots and it is ironic that the country station was the first to be axed from the railway when retrenchment became rampant in the1960s.

2
MARSHALLING YARDS

Until the year of the General Strike in 1926 the railways had enjoyed a monopoly in the field of transport; by statute they were obliged to convey everything a consigner liked to offer, be it elephant or maggot, tree timber or pigeons, coal, or animal feed.

Over the years, with industrial growth engendered by the coming of the railway, a large number of small sidings concentrations had sprung up at the junction end of a cul-de-sac branch, or to serve private sidings in rural and urban areas, or as feeders to goods terminal depots. But these were not marshalling yards in the strict sense. The true marshalling yard is one divorced from the foregoing, being located on a line of route with converging and diverging lines and, in this category on the LNER and Eastern Region came Ferme Park/Holloway, Peterborough, Doncaster, Ardsley/ Wrenthorpe, Colwick, Wath, Mansfield, Guide Bridge (Dewsnap), Dunford, Mottram, Annesley, Woodford, Neasden, Lincoln, Tinsley, Whitemoor, Temple Mills, with York and Healey Mills on the North Eastern and Toton on the former Midland route.

Just as the smaller concentrations grew with local industrial expansion, the larger marshalling yards were located at convenient geographical centres, each some 60 or 70 miles apart, and suitable staging points for the servicing of steam locomotives, for the relief of train crews, for men to lodge or work 'out-and-home' within a shift, and to meet the chief mechanical engineer's requirements for wagon examination which was a vital need in the old days of wagons with grease axleboxes. The former Great Northern and the Great Eastern and Great Central railways each had a progression of marshalling yards along the principal routes.

Annesley-Woodford-Neasden

Annesley,like Colwick, was strategically located at the southern borders of the Erewash and Leen Valleys coalfields but while Colwick gathered and marshalled traffic for a wide spectrum of routes and destinations, Annesley's work consisted predominantly of shunting out the 'shorts', ie, for local stations, for the Great Western via Banbury, for Woodford locals, for Neasden which included Marylebone Goods and the Southern (LSW) via Feltham, all of which was staged to Woodford for changing locomotives and crews. The up yard at Annesley was gravitational into 19 sidings and in 1960 there were 80 arrivals and 64 booked departures a day. The departures included the highly

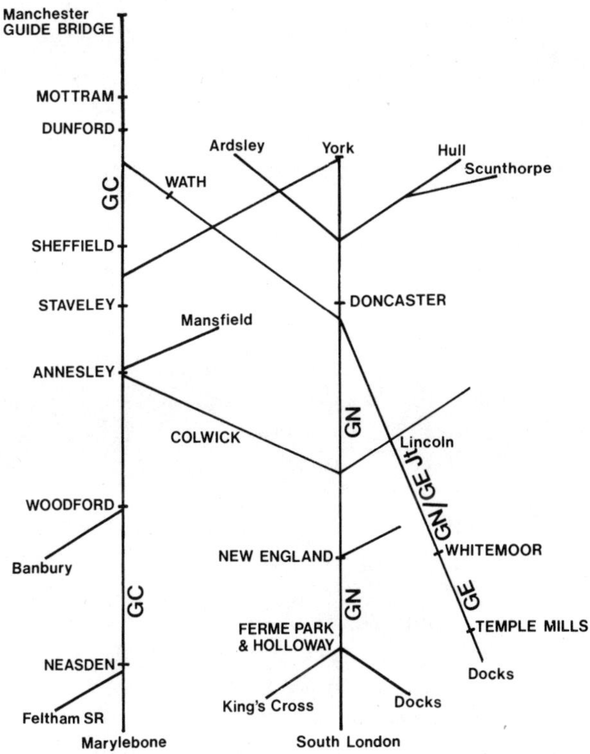

The marshalling yard network on the GC and GN routes, also those providing traffic to the GN & GE joint line to Doncaster from the GE area.

successful and incomparable 'Annesley-Woodford Out-and-Home' service of 31 trains introduced in 1947. The down yard was flat with 18 sidings, having 47 booked arrivals and 58 departures.

Woodford's role was considerably extended during the second world war and after by the opening up of the Stratford and Midland Junction Railway from Woodford to Byfield and Broom for the night shift for the movement of steel trains to South Wales in order to ease the pressure on Banbury. Traffic was heavier to and from the Royal Ordnance Depot at Bicester via Calvert, for the Metropolitan & GC Joint Line via Quainton Road and Aylesbury, and for the GW and GC Joint Line via Princes Risborough and High Wycombe. To accommodate this extra traffic one up and one extra down yard had to be provided in September 1943. In the 1950s the daily average of wagons shunted was:

	Up	*Down*
Annesley	1172	584
Woodford	1300	1497
Neasden	691	933

The difference in the yard turnover between Woodford down and Annesley down was because most trains of empty wagons from Woodford by-passed Annesley on the main line and proceeded straight to collieries. The former Great Central line from Pilsley to Marylebone was transferred from the Eastern Operating Area to the London Midland Region's operation on 1 February 1958. From then there was a general decline of traffic both on the GC route and the Midland route via Chesterfield which enabled the through GC trains to be transferred to the Midland line in June 1965, and this was followed by the complete closure of the spine of the Great Central route in 1966.

A fallacy, seemingly common to the uninitiated, to journalists of the popular press, the pro-road lobbyists and, unfortunately many in Whitehall, was that every wagon went through every marshalling yard on its journey. This has been far from so; a freight planner's maxim was always to secure a full or nearly full train load for conveyance over the longest distance before being broken up for local distribution.

Colwick Sidings

In the railway's heyday of maximum traffic and steam traction the bird's eye view, impression of a large railway concentration could only be one of considerable awe. And so it was, seen from the upstairs windows of the yardmaster's office in a vast area of what was known as Colwick Sidings as distinct from Colwick Marshalling Yard; it seemed to stretch far and wide with scores of sidings full of wagons forming a conglomerate too difficult to take in. Train engines and shunting engines were everywhere and there was a constant grinding of flange on rails, hissing steam, and the banging and clanging of wagons being shunted.

The existence of Colwick Yard came about through the Great Northern's take-over of the Ambergate, Nottingham, Boston & Eastern Junction Railway which used the Midland Company's station in Nottingham's Carrington Street but, after a squabble, built its own terminal at London Road (later known as the Low Level). Having got this base in Midland Railway territory the Great Northern aimed to break the Midland's monopoly in the transport of coal from the Notts/Derbyshire coalfield to London and this it did in 1872 with the authorisation of an extension from the then Colwick farmland to Pinxton, Derby, Ilkeston and even Burton-on-Trent. Tapping new territory demanded facilities for the concentration and sorting, and the choice of Colwick was ideal both geographically and topographically. The approach to the yard was immediately after the GN line crossed the River Trent onto an embankment, the connection coming under the control of a new signalbox named Rectory Junction. The western approach to the yard was immediately after crossing over the Midland Company's Lincoln-Nottingham line. Thus the two high-level approaches enabled wagons to gravitate down appropriate gradients into fans of sidings at lower levels with the westbound exit from Locomotive Junction and eastbound exit from Carlton Fields signalbox. These two signalboxes were in the middle of the complex, and adjacent to the administrative block; the latter was cheek-by-jowl with the old farmhouse which had become unenviably the Yardmaster's residence – the very antithesis of an estate-agent's 'desirable residence' situated with running lines, the yard and locomotive depot on one side, and a swamp being filled with spent ballast on the other. The growth of traffic, particularly from the Erewash Valley called for an expansion of the sidings in 1889 and this was done by laying in aother six reception sidings and another fan of sorting sidings at Colwick North Junction; this was called No 5 Yard or 'New Bank', for coal trains and the original six receptions with accompanying fan became No 4 Yard or 'Old Bank' for goods trains.

Of the six signalboxes Colwick North Junction was the largest. This was a structure part-stilted on the side of an embankment and containing an impressive shelf of block signalling instruments below which was a row of signal, point and lockbar levers, one hundred in all. It was quite a route march for the signalman who was graded Special Class, going from one end to the other, pulling out or pushing back levers. In the centre background was a desk and battery of local telephones and telegraph instruments, the latter giving direct communication with points as far away as Grantham, Derby, Stafford, Uttoxeter, Shirebrook and a host of local places. It was the job of the telegraphist/train register boy to look after these and interpret all the block-bell signalling codes given and received by the signalman and record them in the register accurately and meticulously as, in case of mishap, accident or bad delay, such records could be a vital exhibit at a local investigation or Ministry enquiry.

Colwick North was a junction controlling three main line routes, 12 reception lines into Nos 4 and 5 up marshalling yards known as the Goods Bank (1878/79) and the Coal Bank (1889) respectively, the exit from No 3 down goods yard, the engineer's depot and about three-quarters of the engine movements into and out of the locomotive depot. Trains were always 'in section' and the run-of-the-mill working was quickened by the two nightly express goods trains between King's Cross and Manchester (Deansgate) and their return counterparts. One down and one up was routed via Codnor Park (Brinsley Junction) the Midland route and Chinley and the other one up and one down passed via Annesley GC, Woodhead tunnel and Godley. Other 'liveliers' were the 6.15pm beer train from Burton-on-Trent to Newcastle, and the nightly milk train from Staffordshire/Derbyshire stations to Finsbury Park and King's Cross.

The movements of trains and engines were so intensive as to border on bewilderment – a dream come true for any train spotter. As an example, each morning trains of empty coal wagons for the collieries would be coupled together and run in tandem from Colwick North to Nutbrook Junction and Eastwood on the Derby line and to Hucknall on the Leen Valley. This was the only known locality to adopt this method of tandem working although there are many authorisations for double train working with both engines at the front.

There was a passenger service from Nottingham Victoria to Shirebrook North and Chesterfield via Gedling and, also, three times a day in each direction miners' trains ran between Nottingham (London Road Low Level) and the collieries in the Leen Valley, some of which like Gedling and Teversall, had their own platforms. The coaches on these miners' 'Paddy Mail' trains were spartan with hard wooden seats and minus leather window straps. A tremendous change has taken place over the years as most collieries now have their own commodious car parks.

The yardmaster's office block was large and consisted of private, administrative, inspectors and telegraph offices plus the necessary accommodation for some 300 goods guards, lamproom, and number-takers attached to the Railway Clearing House. The telegraph office had a switchboard for local telephone contracts, and a battery of telegraph circuits stretching as far afield as Grantham, Peterborough, King's Cross, Melton Mowbray, Market Harborough and Rugby in the south, to Newark, Retford and Doncaster in the north, to Ilkeston, Derby, Egginton, Uttoxeter and Stafford in the west, as well as such intermediate local stations as Shirebrook, Pinxton, Kimberley, and the like,plus, of course Nottingham Victoria, London Road, and the Racecourse.

In its busiest days the yardmaster at Colwick was named Edward Lee, a martinet of the old school and a disciplinarian who insisted on being given a good dosage of 'Sir'. He was a non-smoker who jumped on anyone caught indulging or infringing, for smoking on duty was prohibited by the then Rule Book. He was known to call over the coals any transgressor whether or not one of his own staff, or a man belonging to the engineer's department or the locomotive department. On one occasion a goods train was leaving the marshalling yard climbing up the bank towards Rectory Junction to join the main line en route for Peterborough. The guard was standing on the rear verandah of his brake-van, quietly, happily, peacefully, surveying the receding marshalling yard behind him but he had a pipe in his mouth and was smoking! The next day he was on the carpet.

The yardmaster's post was the senior one in the Nottingham GNR district, next to the district superintendent himself, with a staff of about 500 on a salary of £450 a year and dealing with a throughput of around 6,000 wagons a day. The yardmaster never relished the setting up of a district train and traffic control office by the Great Northern at Nottingham in 1911 and which gradually robbed him of status, initiative, scope, and it could be said it knocked him off his perch.

Another local 'dignitary' of the old school in the depot's heyday was the district locomotive superintendent who supervised the running sheds and maintenance of some 400 locomotives of one kind or another, the drivers, firemen, craftsmen, labourers, as well as being responsible for the satisfactory performance of his engines when working trains. He was also in charge of several smaller out-depots and was responsible for breakdown trains/cranes and their working at the scene of any derailment or accident. One particular incumbent was rather dapper, impeccably turned out, habitually wearing spats, yellow gloves and a bowler hat which seemed a little incongruous in such dirty, oily, smoky surroundings. At lunch time on a wet day he would commandeer one of his internal shed engines to run him up towards Gedling past Colwick North Junction signalbox, where at a bridge which crosses the Burton Joyce Road he would drop off and go down to his residence at the foot of the embankment. Down this embankment was a sleepered channel which terminated in or near the coal-house to facilitate the discharge by shovel and gravitation of a wagon of coal standing temporarily on the running line above, the coal having been purchased from one of the local collieries.

Trains of empty coal wagons, privately-owned by collieries

or coal factors, came down from Peterborough (New England) as well as from Whitemoor, the former often consisting of 80 wagons in a train. They were specially signalled and telegraphed forward as 'EW', and were twice as long as the No 1 Yard reception lines at Rectory Junction where they were gravitated into the 31 sidings which had their outlet at Locomotive Junction. No 2 Down Yard (dead-ended) alongside was used for ironstone traffic from Eaton, Eastwell, Denton, Caythorpe and Highdyke, for Ilkeston (Stanton) and Staveley. No 3 Down Yard alongside the locomotive depot was also dead-ended and accommodated the goods and vegetable traffic for Nottingham, Derby, Ilkeston, Burton, Annesley, Uttoxeter and North Staffordshire and the nightly express freights from King's Cross to Sheffield and Manchester Deansgate. Nos 4 and 5 Up Yards each had six reception sidings at Colwick North Junction, the former for goods and the latter for coal. In these two yards saddle tanks propelled the trains out of the reception roads at a steady pace as the foreman shunter, using his shunting pole, uncoupled wagons which rolled down the gradient of 1 in 95, then at 1 in 65 into the capacious sorting sidings controlled by assistant shunters known as 'chasers' who applied the brake lever or levers; this was done with such expertise that they could position the brake-stick and then lean on it and have a balanced but dangerous ride before coming to rest or nearly to rest on the level. The points from the reception sidings into the sorting sidings were controlled by pointsmen in the respective No 4 and No 5 elevated point-boxes according to the code of handsignals by day, handlamp by night, given by the foreman shunter as he made each 'cut'; each signal was acknowledged by the pointsman by hand or light.

The mere fact that there were 12 reception sidings indicated the very heavy traffic sorted for the district and for the more distant yards like Newark, Doncaster, Boston Docks, Grimsby, York, Peterborough, Hornsey, King's Cross, Whitemoor for East Anglia, Grantham, Leicester, Welham, Market Harborough, Northampton and Willesden.

It is difficult to appreciate why the marshalling yard and locomotive depots (the LNW had its own as well as the GN) were called 'Colwick' when the surrounding area of streets was called Netherfield, although the GNR passenger station was named Netherfield & Colwick. Netherfield itself consisted more or less of one main street stretching from the Midland station on the Nottingham-Lincoln line and known as Carlton & Netherfield, to the GNR locomotive depot, yardmaster's office in Netherfield Lane and on to the LNW locomotive depot and the 'LNW Terrace', a long row of terraced cottages on two sides of the depot itself which were built for LNWR men. The terrace was drab and most unprepossessing in appearance, situated with engines belching out smoke on one side and a view looking out on to a swamp being filled in with spent ballast on the other – and the whole atmosphere full of foggy sulphur. From this one long street, Netherfield Lane, intersected by a level-crossing of that name on the Netherfield-Gedling curve, streets went off like ribs from a breast bone, streets which were largely of 'Coronation Street' vintage and decidedly depressing. The terraced houses in these streets, with the occasional side-entry, had no indoor sanitation; it was a 'privy' in the back-yard which was cleared by the council's nightsoil men who started their rounds on a Sunday evening when people were still around in their best clothes. Apart from the Co-operative Society all the local shops seemed to be converted front rooms. Down these streets and down the side or back entries if necessary, the railway callers-up or knockers-up would make their rounds all through the night hours, waking up drivers, firemen and guards with a shout, a bang, some gravel on a window-pane, or by a tap with a clothes-prop left conveniently and purposefully near. Trainmen like those the world over were not on regular hours; they had to take duty at such unearthly odd times, say 2.33, or 3.17, or 4.31. The signing-on time was geared to the departure time of a freight train, due allowance being made for reading up-to-date notices, preparing lamps and equipment, and walking round the train on both sides to see that everything was alright for the journey.

Wherever railways were developed in the 1850s, 1860s, or 1870s it was inevitable that houses had to be provided and so there grew up a community with mutual interests. This is evident in so many railway towns – Doncaster, Darlington, Crewe, Swindon, York – and so in Netherfield. It continued so in the 1920s. The eight-hour day had come in but wages were generally low, yet pride was high and there was a general feeling of well-being culled no doubt from job satisfaction, home, and the allotment, the produce from which was always a good talking point. Train crews were not pegged down to one spot; they could have worked to, say, Doncaster one week, and the next to Stafford, or Peterborough or elsewhere, which gave a variety – and not only variety, as on some jobs it would be possible to bring home a dozen fresh farm eggs, a Stilton cheese, a rabbit or hare, or a pheasant which had misguidedly flown on to the front of an engine and been killed! Wages at that time were £2.10s 0d (£2.50 a week for a porter, 'chaser', or similar grade, rising up grade by grade by promotion or length of service to a maximum of £3.15s 0d (£3.75) which later was the rate for the special class signalmen at Colwick North Junction signalbox. Clerks received £80 per annum at the age of 18 rising by £10 a year to £200 by the age of 31. Promotion before or after then was very uncertain.

In 1916 a trading estate known as the Nottingham-Colwick Estates was established on the banks of the River Trent alongside the Grantham-Nottingham railway and adjacent to the marshalling yard. Although the lines serving the estate were officially the Colwick Estates Light Railway they were of standard gauge and served by both LNER and LMS through a short connection at Colwick East Junction leading into four exchange sidings with a capacity of about 220 wagons. A single line operated by the LNER served various firms' premises and these included oil, concrete, gravel, chicken food, and a sugar beet factory which not only received huge tonnages of sugar beet but also large cargoes of sugar cane from the West Indies through the Humber ports. The sugar beet factory was capable of dealing with about 200 wagons daily and during the beet harvesting season, known in such circles as 'the campaign', which usually extended from mid-September to mid-January a good deal of regulated movement was called for, as for example to the 18 British Sugar Corporation factories. The Colwick factory had the largest intake of wagons – during the 1959–1960 campaign no fewer than 15,102. All this, and much more, alas, was lost from rail to road by the Beeching cuts in the 1960s. Although the Colwick Estates Light Railway was of sub-main line standard it could boast of being a royal road, for on the visit of King George V in 1928 the royal train was run right up to the factory entrance.

In the holiday month of July 1925, the year before the notorious general strike, the throughput at Colwick Yard ("wagons detached" being the yardstick) was at its zenith being no fewer than 147,913, say 5,700 a day, a figure never to be seen again in a comparable period, for the general strike a year later was the thin end of the wedge which began the decline of freight traffic by rail; there were plenty of ex-Army surplus vehicles around and men with entreprenurial skills to exploit the situation and make a dent in carryings by rail. The effect on Colwick is demonstrated by the following figures of wagons handled:

July 1925	147,913
July 1957	89,815
July 1958	76,618
July 1959	68,653
July 1960	50,297
July 1961	55,979.

Meanhile, of course,the second world war gave a temporary boost but the Locomen's strike in 1955 gave the railways another blow. And so it went on, hastened by changes in traffic flow and reductions of coal use caused by the Clean Air Act, domestic and many industrial users of coal going over to oil, electricity and, later, to gas, the elimination of dual-served collieries by all traffic being turned out on the Midland side, the abrogation of road vehicle licensing for the conveyance of goods, permitting a free-for-all while the railway was under statutory obligation to convey everything tendered whether profitable or not and, finally, by the Beeching cuts which stripped the railway trunk services of its feeder traffics from branch lines and hitherto important country towns. The coup-de-grace on Colwick followed the implementation of the Beeching-inspired '1967 Freight Train Plan' when the established service was cut to shreds, completely ignoring traders' requirements with an untenable policy 'Traders will have to accept two-day transits', a theory which had always been unacceptable to traditional railwaymen. The Beeching dictum that the basis of freight operation should be in train loads and not wagon loads, discounting the fundamental fact that 'X' number of wagons comprised a train, consequently the switch to road haulage reduced the number of wagons and the number of trains over all major routes. And so, the Midland main lines from Rotherham, Sheffield, Chesterfield and Derby, to Burton-on-Trent, Toton, Nottingham, Leicester and Birmingham, hitherto accommodating trains to the fullest extent in each direction through the Chesterfield-Clay Cross bottleneck could within a year or two be able to accept the residue of traffic off the Great Central main line and that trickle which had circulated via Colwick following the London Midland's decision not to repair but to close Mapperley tunnel thus effectively cutting off the old Great Northern's 1875 thrust westwards. The regional boundary between the Eastern and the London Midland having been moved back from Netherfield and Gedling to Aslockton on the Grantham line put Colwick Yard entirely within the LM Region and, totally denuded of traffic and life-expired, it was closed in 1970, aged 91. The saga of antagonism between the Midland and the Great Northern inthe Nottingham area would not be complete without turning the clock back to the 1950s when Dennison of the GN wrote to Ellis of the Midland in a letter dated 16 August 1852:

> . . . an earnest attempt ought to be made to unite the GN and the Midland railways upon the principle of complete amalgamation. They compete with each other in the south and in the north, and they cross each other at two or three important points; there are double stations at several towns and duplicate trains run where single ones would serve the public equally well. A very large annual expenditure would therfore be saved, and an amalgamation is so natural from peculiar circumstances, and so inevitable that, I apprehend, no Parliamentary objection would be offered.

This far-sighted statesmanlike, businesslike, approach was turned down by the Midland, preferring an alliance with the London & North Western. What an enormous capital expenditure would have been saved if it had been taken up, for many branch lines were duplicated in the same valleys, and many duplicate connections with collieries would never have come about.

Peterborough

The first railways into Peterborough were the Eastern Counties and the London & North Western in 1845 when they formed an end-on junction at the Eastern Counties station which later became known as Peterborough East. The Midland Railway branch from Saxby and Stamford followed in 1846, the Great Northern's 'Loop Line' from Peterborough to Spalding, Boston and Lincoln in 1848 in anticipation of the GN's completion of the London-Peterborough section in 1850. The opening of the GN line from Peterborough (Werrington Junction) to Retford came in 1852 and so facilitated the GN's take-over of the Ambergate Railway from Grantham to Colwick and Nottingham in 1850, and the laying in of a physical connection at Barrowby Road. As with Colwick, so with Peterborough. The growth of coal traffic, greatly in demand in London and East Anglia, sparked the fierce competition between the Midland and the Great Northern Companies.

The Midland had already established a small marshalling yard at Wisbech Sidings with its line running through to connect up with the LNWR and Eastern Counties at Fletton Road Junction. The opening of the routes from Colwick and Doncaster called for a marshalling yard at Peterborough which was located at New England, on the east side of the main running lines; parallel on the west side of those running lines were the Midland running lines and Wisbech Sidings. The increased traffic of a miscellaneous nature over the years, both terminal and through, necessitated the extension of the facilities at New England; the original yards were the Old Coal Yard, single-ended with no reception tracks, and the Old Down Empties. Then came the East Coal Yard with four reception lines and gravity shunting, the West Empties with five receptions and 22 sidings, Spital and Westwood. All were flat except two. The Old Coal and the Old Empties were later used for through loads not requiring shunting but calling in for locomotive and crewing requirements or pathing.

In 1893 the M&GN from the Norfolk coast, King's Lynn, and Wisbech came into Peterborough from the east, its tracks flying over everything else to drop down on the west side of the Midland lines, joining the latter at Wisbech Junction and the Great Northern at Westwood Junction. The M&GN had no sidings of its own, attaching, detaching, terminating and starting its freight trains at Wisbech Sidings (Midland) or Westwood (Great Northern). The Eastern Counties (later Great Eastern) had its own groups of sidings at Peterborough East, Stanground and Whittlesey and leased some accommodation at Peterborough East to the London & North Western Company. Inevitably there was a good deal of transfer and trip working from one company's yard to another and heavy delays were occasioned to GN trips from New England to Peterborough East because the Midland held the key at Nene Junction and ruthlessly carried out a policy of two Midland to one Great Northern, a system which persisted until the whole operational complex came under Eastern Region control after Nationalisation.

The main marshalling yard was New England. It was the natural strategic staging point, but not necessarily the shunting point, governed by distance, steam locomotive and train crew considerations, and wagon examination (axleboxes, couplings, etc,) for trains between Nottinghamshire, Lincolnshire, and Yorkshire areas and London, as well as being the gathering ground for much agricultural produce originating in Lincolnshire, north Cambridgeshire, Huntingdonshire and Bedfordshire, and the very extensive brick traffic from the nearby kilns at Fletton. There were separate Up Goods and Up Coal Arrival lines from WerringtonJunction leading into the respective yards at New England North. There were separate yards for the nightly express goods trains in the up direction at Spital Yard and in the down direction at Westwood. Freight trains from the M&GN line had to cross over the Midland line at Wisbech Junction and traverse the 8 chain connection to Westwood Junction where they backed into the yard arrival line from the up GN main line. The flat Westwood Yard with buffer stops at the south end was always noted for the alacrity with which it accepted and shunted a quick succession of fruit and vegetable trains from the M&GN together with those from Sandy, Biggleswade, Ramsey, Holme, etc, with several individual wagons or 'cuts' rolling off at the same time with the hand points being smartly pulled over between each cut when only yards apart. It was an operation in which the shunting staff took immense pride. But then, pride in their work was common in those days!

Traffic for London, King's Cross, the Docks, and the Southern Railway was sectionalised at New England into various categories to facilitate work at Hornsey (Ferme Park) and Holloway (East Goods), separations being made for the SE&C, the LB&SC, the LSWR and coal for the Highbury Vale depot and High Barnet and Edgware branches then owned by the Great Northern/London & North Eastern.

Traffic for important roadside stations like Biggleswade, St Neots, or Sandy, was very heavy and every morning three long freight trains left New England to serve the stations between Peterborough and Hitchin, each being manned by two guards because of the length of the trains. Alas, no more! Every night two trains of potatoes left New England for King's Cross Goods potato market; one started back at Algarkirk and the other from Long Sutton. Vast quantities of vegetables were despatched on afternoon and evening express freight trains, so affording arrivals in the Midlands, Merseyside, the North-East, and Scotland in time for early morning markets.

The haulage of coal over long distances to municipal power stations and gasworks deceased with the nationalisation of those concerns and the setting up of huge modern power stations near the seat of coal production. The railways turned away gradually from steam power to diesel and electric. The opening of the mechanised marshalling yard at Whitemoor, north of March, in 1928, had enabled more trains to be channelled via the more direct route through Spalding instead of through the Peterborough complex. Competition from the road hauliers from the general strike period of 1926 for the conveyance of vegetables and other perishables became more intensified as the years passed, biting into merchandise traffic generally. The Beeching Axe fell on the M&GN Line on 2 March 1959 and the Peterborough-Rugby and Peterborough-Northampton lines were closed on 6 June 1966. Many other individual stations followed. The Beeching Axe, however, had been pre-empted by the LNER insofar as it had a 'Post-war Development Plan for Peterborough' in existence. Each company had its own marshalling yard facility, goods depots, locomotive sheds, booking offices, personnel and such condi-

Above The north end of Peterborough, with Westwood Junction signalbox on the left controlling the GN main line and the crossovers to the Midland line at Wisbech Junction on the far left. To the right is Eastfield leading to the New England yards and engine shed. (L&GRP/David & Charles); *Below* Ferme Park yard. (L&GRP/David & Charles)

Above A Great Northern general through freight at Radcliffe on the Nottingham–Grantham line in the 1920s. (L&GRP/David & Charles); *Below* Coal train on the Great Central hauled by 2–8–0 No 6320 near Leicester in the 1930s. (L&GRP/David & Charles)

tions were over-ripe for rationalisation. Although under discussion in 1944 and later, the London Midland & Scottish showed its usual reluctance and this, coupled with the London & North Eastern headquarters seemingly being more concerned about station façades than track re-alignments and re-organised marshalling yards, meant that the situation became sterile for a long time. However, the long gestation of a rational redevelopment programme began to take shape in the 1950s and early 1960s. The main features of this were the laying in of running junctions between the GN main line and Midland main line between Werrington Junction and New England North to permit an interchange of trains. Thus LMR trains destined for Wisbech Sidings would be dealt with in New England and with the residue of traffic being diverted at source via Whitemoor the way was paved for the closure of the Peterborough East and Wisbech Sidings yards in April 1953.

During the later years of the second world war consideration was being given to a post-war development plan for Peterborough North station, which had only one platform face for through southbound up trains but two platform faces for the northbound down trains. There were occasions in the up direction when a Midland passenger train could not be timed over the GN main line from Westwood Junction to call at the up platform face before proceeding to the East station so that trains had to pass the North station on the through Midland line leaving any passengers for the North and GN connections to make their own way back from the East. Apart from this disability there was the sharp reverse curve through the North station with a 20mph speed restriction at each end of the platforms, a relic of the 1850 construction. Although a remodelling of the station and through lines with a new island platform situated on the west side to service the ex-Midland and M&GN trains and so relieve the GN up main between Westwood Junction and Crescent Junction was urged in 1949, it was not until 1973 that it was done under a £2½ million scheme to permit the new High Speed expresses to travel through at up to 100mph.

Until 1913 there was another relic of the 1850s in the form of a road level-crossing at Crescent Junction at the south end of the GN (North) station and which spanned all GN an

Midland tracks; it was a great relief when a road overbridge was constructed. Until 1924 there was only one up and one down line over the River Nene between Fletton Junction and Crescent Junction and a new girder bridge enabled the tracks to be quadrupled and proved a boon to Crescent Junction turning out freight trains with smaller margins to Spital Junction for the New England yard.

Southbound trains were normally turned out of New England yard at Spital Junction across on to the up goods line on the west or down side of the North station and back again to the up main or up goods line at Crescent Junction, so fouling the down main twice; on occasions they were turned out at Spital Junction to the up main if there was the possibility of the up main platform face being free of passenger trains. During the second world war this hazard was eased somewhat by a siding between Spital Junction and Peterborough North signalboxes being upgraded as a running line and duly signalled to enable a freight to creep forward to Peterborough North and dash through the the platform line as soon as a margin between passenger trains permitted it.

Peterborough East station was also plagued with a road level crossing from the mid-19th century at Fletton Road Junction just west of the station platform, until 1934 when it was closed on the construction of an overbridge to take the road traffic. With the later closure of the LNW line the junction and signalbox were closed.

In the mid-1950s plans were afoot to modernise the New England yard layout by reducing the six yards to three with a saving of 54 men and seven locomotives but rationalisation took priority over modernisation. Loss of traffic for one reason or another and the general run-down has proved this was the right course to take, but it is nevertheless ironic that after two level crossings had been closed, lines quadrupled over the River Nene, interchange junctions laid in, curved tracks through the North station straightened out for high-speed running, and modern signalling from one power signalbox instead of many manual boxes, the end result is less traffic, both passenger and freight, and fewer trains! And how much more has it cost to provide a new station, something which looks as cheap and mean as a bus shelter?

It was in the early 1920s when the author first knew Peterborough, still at that time in all respects a bustling railway centre, being the exchange point for both passengers and freight between five administrations. Although the Railways Act operative from 1 January 1923 grouped the railways, all it had done for several years was to put the GN and the GE into the London & North Eastern Group and the Midland and the London & North Western into the London Midland & Scottish Group, leaving the Midland & Great Northern Joint to continue working as a separate entity under the Resident Manager at King's Lynn until later absorbed by the Cambridge LNER district and the Norwich LNER district.

The fact that there were two routes from Peterborough to North Cambridgeshire and the Norfolk Coast not infrequently caused passengers some difficulty, as they found themselves at Peterborough holding a ticket routed via the M&GN when the last train had gone, or vice versa the GE route via March, and thus had to be given a written permit to travel via the alternative route to the stations mainly affected, such as Wisbech, Lynn, Cromer, Sheringham, Norwich, Yarmouth and Lowestoft. Passenger ticket and freight routing presented difficulties in accountency and operation until sorted out by rationalisation which closed the M&GN Joint Line almost completely in March 1959 finished it off.

The individuality of each railway company created a hotchpotch of engine running sheds just as happened in the way of marshalling yards. The Great Northern had its shed at New England, the Midland at Spital Bridge and London & North Western at Woodstone and the Great Eastern at Peterborough East. In 1932 the Midland and the L&NW came together at Spital Bridge and in 1939 the GE shed was closed and the workings transferred to March. The M&GN engines and men were finally located at New England in 1939 and remained there until the line closure 20 years later. In their 'prime' the sheds at Peterborough serviced a combined total of 350 or so engines. New England shed, as the largest, is the only one to survive.

Hornsey and Holloway

The London marshalling yards at Hornsey and Holloway formed the last stage of the Doncaster-Colwick-New England-London freight run; the siting of those yards was more fortuitous than strategic but the distances between each were somewhat equidistant to suit the steam engine servicing, train crews' hours of duty, and examination of wagons, bearing in mind the distance limits which the latter had when fitted with grease axleboxes as the majority were at that time.

Holloway yard was immediately south of Seven Sisters station (which was later re-named Finsbury Park) and consisted of East Goods Yard on the up or southbound, and Clarence Yard on the down or northbound. Nearby on the lower level of the Finsbury Park – Canonbury line was the Highbury Vale Yard. Each yard was small and quite incapable of expansion for topographical reasons so that traffic had to be quickly processed. This was not easy in view of the prohibited hours for the working of freight trains to the London Docks over the North London Line via Canonbury and to the yards and depots south of the Thames via the Metropolitan line and Blackfriars, which route was first opened with a simple connection between the Metropolitan and the Great Northern at King's Cross on 1 October 1863 but improved on 1 March 1868 by slewing and constructing separate tracks known as the Widened Lines to accommodate the Moorgate passenger services from and to the Great Northern and from and to the Midland Railway.

Hornsey Yard comprised Ferme Park Up and Ferme Park Down with a double line viaduct connecting each other over the main passenger lines. The Up yard was on a rising gradient of 1 in 280 running contrary to that required for shunting at the north end in the direction of traffic flow. In a general sense the Up Yard dealt all the heavy coal trains and rough goods, leaving East Goods to specialise more on the express goods, fish and perishables. Ferme Park Down dealt mainly with empty wagons and rough goods which were not suitable for the express freight trains concentrated on Clarence Yard, which acted as a sifter of traffic from the London Docks and south of the Thames, extracting that traffic which was suitable for combining with the express goods trains which started back from King's Cross Goods and called at Clarence Yard ten minutes later.

In July 1925 Holloway Yards detached 50,158 wagons; in the comparable periods of later years the figures were:

1958	22,878
1959	18,935
1960	1,346
1961	1,298

Ferme Park Yards detached 71,282 wagons in July 1925 and in the comparable later periods:

1958	30,683
1959	33,692
1960	46,054
1961	46,286

Although traffic movement had been influenced by the dock strike in 1955, the Suez crisis in 1957, and a trade depression in 1958, the decrease at Holloway and the increase at Ferme Park during the four years 1958–61 was due to a planned rationalisation, in which both East Goods and Clarence Yard were progressively closed, the King's Cross/Clarence Yard freight trains to the north reorganised, leaving Ferme Park to deal with the residue. This was highly successful and paved the way for Clarence Yard to become the site for the new Finsbury Park diesel depot.

Up or southbound traffic dealt with at Ferme Park and East Goods had for many years had a high proportion of coal. With the opening of the Metropolitan Widened Lines between King's Cross Metropolitan Junction, Farringdon Street, Holborn, and Blackfriars, the Great Northern had built up a service to the South Eastern & Chatham, the London, Brighton & South Coast, and the London & South Western Railways, to yards at Battersea, Hither Green, Herne Hill, Bricklayers Arms, New Cross Gate and Feltham and to the ex-GNR coal depots at Elephant & Castle and Brockley Lane; in the latter days of the 1890s as many as five trains daily were run to the former. The heaviest period of train working was during the second world war but the volume began to decline thereafter which caused an inter-company meeting between the LNER, LMS, Southern, and London Transport to be called to prune the timetabled service then at 65 trains daily in each direction to a level of realism. Valuable as the Widened Lines route was (at one time passenger trains ran from Victoria to King's Cross and some GN suburban stations) it was restrictive by the the prohibited periods of approximately four hours during the morning passenger peak and another four hours in the evening. Engine route availability was restricted to Class J52 0–6–0 saddle-tanks, N1 0–6–2 tanks and J50 0–6–0 tanks because of weight limits on Southern Railway bridges. The train length was restricted to 32 wagons standing clear at Farringdon Street, at which point the LNER provided a banking engine to assist trains up Snow Hill. Each night two fish trains left East Goods for Clapham Junction and London Bridge. Today not a single freight train passes over the Widened Lines/Holborn Viaduct route, the line being reserved just for London Midland Region passenger trains into Moorgate, with the link to the Southern lines long since severed. However, moves are in hand to restore this north-south link.

Other freight trains ran from Ferme Park and East Goods to Feltham (LSWR) and to Acton (GW) via Canonbury and the North London Line, also from Ferme Park to Acton via the Harringay Curve and the Tottenham & Hampstead Joint Line after the war-time connection from the GN line became available. Freight trains also ran to Stratford, Mint Street, Poplar Dock and Victoria Dock, bringing back traffic to connect with the northbound express freights from Clarence Yard. A high proportion of traffic for King's Cross Goods depot went straight through on express freight – fish, meat and vegetable trains being followed by the rough and heavy consignments on trips from Ferme Park and/or East Goods, already sectioned in twelve sorts for ready placing and convenience of traders. The smaller yard at Highbury Vale on the lower level of the Canonbury Line adjacent to East Goods Yard was used for marshalling seven trains for Hackney Wick, Bow Common Gasworks, for the 'Northern Branches' of High Barnet, Edgware, Highgate and Alexandra Palace, and became virtually redundant when the London Passenger Transport Board took over jurisdiction and electrified the latter branches (except Alexandra Palace) as extensions of the Underground system.

Doncaster

The Doncaster Marshalling Yard was really a complex of yards covering north/south and east/west flows with transfers to and from each group, a relic of and indicative of the piecemeal growth of no fewer than nine constituent companies running into Doncaster over their own routes or exercising running powers. They were the GN, GC, GE, NE, H&B, Midland, LNW and L&Y, and the South Yorkshire Joint Line (the latter not to be confused with the earlier South Yorkshire Railway which itself was a constituent of the GC). At the time of the Grouping on 1 January 1923, the Great Northern had five yards – Down Decoy, Up Decoy, Down Mineral, Up Mineral, and West Yard, while the Great Central had four yards – Marshgate, Hexthorpe Up, Hexthorpe Down, and Hexthorpe Top Yard; shortly after, the Hexthorpe yardmaster (Paul Jepps) was responsible for all the yards outside the precincts of the passenger station. With the amalgamation the Great Central's district operating office at Doncaster (St James' Bridge) was combined with the Great Northern's district office which moved from Grantham to form a new district command located at St James' Bridge but in later years transferred to Gresley House.

The number of wagons detached in the nine yards during the month of July 1925 was then 138,990 but in the comparable period of July 1961, six yards then making the whole, detached 51,933, a very substantial decrease. Between the two dates, in the immediate post-war period when some semblance of normal trade and traffic movement was being restored, a long hard look was taken at the possibilities of rationalisation; after re-organisation of train and trip working following a re-routing exercise the following Hexthorpe yards were closed: Top, 2 June 1949; Up, 9 October 1950, and Down on 2 July 1951.

The wagon marshalling facilities at Doncaster were no different from Peterborough, and many other centres, being a hotchpotch of individual company efforts to meet the growing needs of the time and with little farsightedness. The Great Central's top Yard at Hexthorpe was an obvious addition being laid down in advance of the Up and Down Yards because there was no land available to expand sideways. As for the Great Northern there was a little more scope. Basically the four yards were Down Decoy, Down Mineral, Up Decoy and Up Mineral. Down Decoy and Up Decoy were flat and dead-ended, the Up Mineral was double-ended and gravitational. At some time before the Grouping the Up Decoy Goods Yard was extended sideways because land was available and that particular section became known as 'New Yard' and was used largely for marshalling trains serving country stations in station order.

In the 1950s the work of Down Decoy Goods and Down Mineral yards was merged when the down Mineral was reconstructed from a dead-ended flat to a double-ended gravitational, the dismantling of a disused overbridge making it possible to lay in some reception lines on a rising gradient. As the Down Mineral was then to be used for all freight traffic it was re-named Belmont and the Down Decoy yard, closed for shunting, was retained for storage. Out of the original nine yards the three remaining in 1965 were dealing with approximately 1,046 (Belmont), 709 (Up Mineral), and 650 (Up Decoy) wagons daily.

Apart from the characteristic run-of-the-mill goods and coal traffic there were many highlights which were not restricted to the long-distance express goods trains which occupied most of the night-shift hours, calling at Decoy Yard – so named because of duck-shooting which took place in the marshes many years before – to detach, attach, and be re-formed for the onward journey. The two fish trains from Hull to King's Cross and south of the Thames, and the two for Banbury, (for the West Country and South Wales, plus vans for Nottingham, Leicester, Oxford, for detaching en route) made a call at Up Decoy if only for engine and train crew change, and to see them depart was a thrilling sight particularly if the renowned driver Joe Duddington (driver of *Mallard* on its 126mph record run) had his hand on a regulator. He had steam in the cylinders and made a get-away quicker than any other driver.

Although the North Eastern had absorbed the Hull & Barnsley Railway like the London & North Western had absorbed the Lancashire & Yorkshire Railway before the 1923 amalgamation little progress had been made towards altering operating practice. For instance, the H&B's way into Doncaster was via Hemsworth Junction which entailed a reversal at that point and this continued until the late 1920s or early 1930s when all Hull traffic was routed via the Stainforth-Thorne Junction route. But until then the H&B caboose with the roof-top look-out was a familiar sight as were trains composed of 100 wagons starting from Down Decoy en route for Hull Docks, threading their way through the numerous block-signalling sections and the Central passenger station before branching off at Marshgate Junction; a single train could thus be 'in section' at several closely-spaced signalboxes at one time.

Much traffic was seasonal. The West Yard was given over to dealing with the heavy passenger-rated fruit passing from Cambridgeshire during a period of about three months, while the goods rated fruit and vegetables were dealt with at Down Decoy. It is difficult to believe or appreciate the quantities and variety of traffic dealt with in days of yore. Every night in the winter a train conveying celery came in from Haxey, Finningley and Park Drain for shunting and despatch in many directions. There was a special train numbered 613 carrying solely green peas and beans gathered fresh from the fields that day in the Doncaster, Goole, Selby area. This left Doncaster Decoy at 8.36pm for King's Cross where it was due at 1.40am but detaching at Holloway (East Goods) 1.05am to 1.30am traffic for immediate transfer on trips to Maidstone, Kew Bridge, Stratford Market and Spitalfields. The green peas in sacks were loaded in cattle wagons withdrawn from normal service, specially cleaned, and used because the open slatted sides provided a current of air and prevented sweating. Also, commencing as a rule in the first week of the New Year a train of rhubarb ran as a nightly express from the West Riding to King's Cross until the end of the season; its working timetable number was 906.

The St Leger Race Meeting was THE event of the year at Doncaster and extensive traffic working adjustments had to be made to cater for the many excursion trains which brought the people in; road transport by coach or private car had not yet taken a hold. The main Central passenger station could not cope with the many extra trains and some had to be farmed out to the old St James' Bridge platforms (a relic of the South Yorkshire Railway) – those from and returning to the West via Rotherham and Sheffield, Chesterfield, and Manchester – while the West Yard alongside the Plant Works and the Central station took on some of the extra trains for the West Riding, Lancashire and the North but in this case passengers had to climb up wooden steps to get in the coaches. The Down Yard at Decoy presented a spectacle which delighted many an enthusiast, man or boy, for there, on the Leger afternoon, lined up side by side, siding by siding, were ten or a dozen sets of coaches each with a 'foreign' engine on the front, waiting for the signal to be off for each to leave in the appropriate order of procession, to fit in with the booked departures of the trains from the Central station for Leeds, Hull, Scarborough, York, Newcastle, Leeds, Bradford, Halifax, Blackpool, and Lancashire connections. According to Sir William Ackworth in his *Railways of England* published in the late 1880s, 'On the Leger Day of 1887 the Great Northern Railway coped with no more than 82 Excursions to the town'.

Railways and Doncaster have been synonymous (like Crewe and Swindon) since 1851 for at that time the Great Northern directors decided to transfer their repair works from Boston. The original 500 or so men affected grew and grew over the years into an establishment of over 5,000 necessitating the building of many houses and also St James' Church and day school for their spiritual and educational wellbeing. The Works or 'Plant' covered an area of 84 acres and had responsibility for the repair and maintenance of some 2,000 locomotives, 2,500 passenger coaches, and 30,000 wagons, quite apart from the construction of stock to existing or new designs. This was the base for a succession of brilliant locomotive superintendents later designated chief mechanical engineers – Stirling, Ivatt and Gresley – who covered the period from 1866 to 1941. Stirling's 4–2–2 in 1898, and Gresley's 4–6–2 in 1922, ensured amongst things that Doncaster had a special place in the railway world. In its turn the railway had a special influence on Doncaster almost to the exclusion of any other major industry, until, by 1926 new collieries had been sunk on the town's periphery – Brodsworth, Bullcroft, Bentley, Rossington, Yorkshire Main, Markham Main, Hatfield Main and, later, Harworth, Firbeck and Thorne – and were beginning to have a growing impact and influence on the life of the town despite each colliery having its own village community and Miners' Welfare Clubs.

Locomotive servicing of trains was done initially by the Great Northern at its Carr running sheds alongside the Up Mineral Yard and convenient for the Central station and Decoy. The Great Eastern Railway maintained its own small shed for some years until entering into an arrangement like the North Eastern, H&B and the L&NW to use the Carr facilities. The Great Central trains were worked by engines from Mexborough or Keadby until the district locomotive superintendent set up a new district based on Carr depot. That depot serviced a stud of 13 Pacifics, a large variety of freight engines and no fewer than 25 freight shunting engines, the latter being manned continuously from the early hours of Monday through to 6.0am on the following Sunday in the various sections of the yard. This stud of locomotives consumed up to 3,000 tons of best steam-raising coal each week.

Whitemoor

The country town of March in north Cambridgeshire was the meeting point of several railway routes serving not only the farming country of East Anglia but also the coalfields of the Midlands and Yorkshire, the docks of London, Harwich and the North East and North West ports. But March Yard was unable to meet all needs and in 1928 what became Britain's biggest marshalling yard was opened at Whitemoor, immediately north of March on the GN & GE Joint Spalding line. It was 2½ miles long and a mile wide, contained 70 miles of track and was soon handling 7,000 wagons a day. Although not the first yard in the country to have power worked points it was the first to be fully mechanised. When trains arrived in the Down Yard receptions or the Up Yard receptions they were checked for damage and mechanical faults and a shunter or 'cutter-up' marked on a cut-card what should be done with each wagon. The train was then propelled to the top of the hump and from there the wagons rolled down into appropriate sidings. Meanwhile the operator in the control tower had received a copy of the 'cut-card' which enabled him to pre-set the electrically-controlled points leading from the hump into the sidings, the settings being stored in a machine until the wagons started to roll down; the first wagon over the points automatically set them for the next cut and this continued until the last wagon on the train had passed over. This fully automated yard with power-operated retarders instead of men 'chasers' applying wagon brakes quickly became a show-piece. Gone was the flag-waving and signalling by handlamp, for the 'cutter-up', the shunting engines and the control tower were all linked by radio, thus ensuring that darkness and fog had practically no effect on communications.

Coal formed a heavy proportion of southbound traffic into the Up Yard and, conversely, empty coal wagons going back to the Nottinghamshire, Derbyshire and Yorkshire collieries were a predominating feature of the throughput in the Down direction, especially when such empty wagons were privately owned and labelled to return to a specific colliery or group of collieries. The requisitioning of those privately-owned wagons by the Government at the start of the second world war so that not only did they become common-user but could be used for traffic other than coal, was a distinct boon to movement through marshalling yards and eased the overall pressure. One benefit was that the wagons could be re-loaded where the coal had been discharged with sugar beet going northwards to the sugar beet factories, notably Colwick.

While freight traffic to and from East Anglia had been largely of a rustic nature, that is connected with agriculture generally, the second world war brought about great changes, particularly in building up and maintaining the British and American Army and Air Force bases. Seven-day freight working became a dire necessity when morning stocks in the Great Northern pipeline averaged 1,848 wagons in 1942 and no less than 2,575 in 1943. Much traffic from the North East via York had to be diverted via Applehurst Junction, Stainforth and Scunthorpe to Immingham, there to be pre-sorted into block loads to by-pass Whitemoor. The cessation of hostilities enabled a more stable pattern of train working to be organised with an ever increasing volume passing on vacuum brake-fitted trains, such development depending on the increasing avilability of brake-fitted wagons. The 8.40pm express freight from Whitemoor to Niddrie (Edinburgh) started the night run of despatches followed by the 8.50pm to Ducie Street (Manchester), 9.3pm Ardsley (West Riding), 9.40pm Colwick, 10.10pm Doncaster, 10.25pm Stockton, 10.40pm New Clee and supplemented during the Cambridgeshire fruit season by earlier departures. In the Up direction express freight trains ran from the West Riding (the 9.30pm Ardsley-Spitalfields and the balancing Down train were long-standing services operating long before Whitemoor opened), Manchester, Doncaster, Hull, Niddrie, Newcastle, York, Spalding, Wakefield. The rationalisation of the Peterborough marshalling yards resulting in the closure of three yards there and the extension of the LMR Toton, etc,–Peterborough trains through to Whitemoor, and the closure of the L&NW line into Peterborough East gave the opportunity for Whitemoor to introduce new through and express freight services to Birmingham, Crewe, etc. Freight services southbound from Whitemoor had an extensive spread to the London terminals, dockland, and throughout East Anglia. When the second world war broke out in 1939 Whitemoor yard had been opened 11 years and it is difficult to imagine how, without it, such a task could have been maintained.

Bradford (in 1933–34)

Just as the country station dealt with a variety of traffics so did the town or city depot but in much larger quantities and doubtless in still greater variety and Bradford can be taken as being very typical of what was done. There were two LNER goods depots in the city, the one nearest the centre being Adolphus Street while the other was at City Road, a little nearer the outskirts on the Halifax-Keighley side. The supervisory side of the two depots was a strange one in a way for although the goods agent had around 100 office staff all were in the commercial department except one who was a a lone operating man. The latter's immediate chief was on the staff of the district office in Leeds and he held the dual position of district inspector and assistant yardmaster at Bradford. He was responsible for all movements whether by hand or wheel at both City Road and Adolphus Street.

Adolphus Street was the main depot of the two and was reputed to be one of the largest goods terminals in the country; it was unusual insofar as the offices were in the original buildings of the first Great Northern passenger station in Bradford and the concourse and platforms were used as bays for handling the small merchandise received or forwarded, being passed from one wagon to another or trollied around. Every consignment was invoiced and hand-written hence the large clerical staff as mechanisation had not come on the scene. The office dealing with the forwarded small or large goods ws known as the 'shipping office'. The volume of traffic was still heavy as many of the merchants had the facility to warehouse on railway property and withdraw raw materials as required, a facility which road competitors could not match.

Adolphus Street depot was sited on the side of a hill between Laisterdyke and St Dunstans or Bowling Junction, and the entrance by rail was via Hammerton Street Junction, which signalbox also controlled the entrance/exit with the locomotive depot. Being on the side of a hill the whole depot was really a series of smaller yards all on different levels and with gradients, but each convenient for dealing with its own more or less specialised traffic. For example, there was the 'Arcade' (the one-time passenger station) dealing with 'smalls', the main yard for general, the others being for the grain shed, the wool shed, the sugar shed, the timber field, and the cattle dock which was next to the cattle market so that there was no movement of cattle through the streets.

Underneath the 'Arcade' in the bowels of the depot was the banana warehouse, for many wagons of bananas were constantly arriving from Avonmouth and Garston Docks, the bananas in their green state to be stored in the warehouse and ripened by heat from gas jets for the retailers' shops.

Wool was very big business and hundreds of wagons containing huge square bales of wool arrived and were passed through the wool shed or held up in warehouse storage awaiting consignees' orders based on mill, trade or market fluctuations. In those days large firms hired warehouse space for wool, sugar and grain and the intake, handling and ultimate despatch was dealt with by the railway gangs, much of it being casual labour – three men working to the instructions of an established railway foreman. There was at this time, 1933 and 1934, a high measure of unemployment and each morning about 8.30 or 9.00 some 15 or 20 men would be waiting at a pre-arranged point in the depot for the assistant yardmaster to select from them, so many gangs of three or four men each, with orders to go to the wool shed or wherever; those who were not engaged for the day would wander back to the unemployment exchange unless they managed to get work elsewhere or hoping for better luck on the morrow. In the cattle dock only one man was employed, for the cattle, not unexpectedly, had legs, and could move themselves, contrary to the rest of the depot's traffic. The man was Tommy Trout and his job was to assist the consignees' drovers to unload the cattle from wagons into pens before being driven off to market or the abbatoir, and to clean the wagons and pens afterwards by shovel, hosepipe and disinfectant. Rationalisation later centred all the Bradford freight traffic on to the ex-L&Y depot at nearby Bridge Street.

In the passing it should be mentioned that an interesting feature of the working at Adolphus Street at the time was that as the cobbled streets approaching the depot were steep, the horse-drawn loaded drays could not manage the climb without assistance. A spare horse was stationed at the bottom of the street and yoked onto the front. Afterwards the horse was replaced by a tractor fitted with a steel breastplate, which assisted by pushing from the rear.

Sheffield and Tinsley

The Transport Act of 1947 and Nationalisation in 1948 brought regional boundary changes in 1950 which consigned the whole of Sheffield to the full administrative control of the Eastern Region. The London & North Eastern and the London, Midland & Scottish companies had a multiplicity of stations, yards, depots, and locomotive sheds, even though a measure of agreed resource pooling had been carried out in the 1930s to avoid wasteful competition. By 1950 some progress had been made with modernisation, including the Manchester–Sheffield–Wath electrification, completed in stages, in 1952, 1954, and 1955.

In 1957 under line management the new Sheffield District was formed, being an amalgamation of parts of the GC in the Lincoln, Manchester, Nottingham and Doncaster districts, and the London Midland & Scottish Rotherham district which stretched from Cudworth and Darfield in the north to Chesterfield inclusive in the south. This brought all traffic movements under one control and the basic requirements for this were one marshalling yard, one freight terminal and one motive power depot. Tinsley was the site chosen for the yard and locomotive depot for maintenance and servicing, and Grimesthorpe as the freight terminal. Topographically Tinsley was not the fully ideal site for the yard but it was the only possible and available site which could provide new junctions laid in to both GC and Midland lines with full exchange and giving access in every direction. Covering 145 acres there were 11 reception roads six of which were wired at 1,500V dc for Manchester-Sheffield-Wath electric services, 53 sidings in the main yard, and 25 sidings in the secondary yard used for local traffic. Then there were five sidings in the express yard used for through express freight trains calling to detach and attach wagons. The yard had a planned capacity for dealing with 4,000 wagons daily making up trains for every direction – Toton, Norton Junction, Washwood Heath, Bristol, Port Talbot, Chaddesden, London (Brent), Healey Mills, Hunslet, Carlisle, York, Glasgow, Birkenhead, Liverpool, Mottram, Hull, Nottingham, Colwick, Worksop, Doncaster, Whitemoor, Lincoln, Peterborough, and King's Cross. The yard had the most modern techniques including Dowty booster/retarders which automatically accelerated slow-running wagons as well as retarding fast-running ones, with reference to a computer-calculated optimum speed.

Nottingham District Control Office; the section controllers are on the right with a track diagram in front of them and pegs to indicate train position and other operating details (*Lincolnshire Echo*)

Controller's nightmare – 1. On the former double track bottleneck through the tunnels between New Barnet and Potters Bar on the otherwise four track section at the southern end of the GN main line, signalmen and controllers had a tough task in keeping slow moving trains out of the way of express passenger trains. *Above* All-stations local from King's Cross to Hatfield. *Below* Special train of empty cattle wagons climbing towards Potter's Bar. (L&GRP/David & Charles)

The provision of this huge marshalling yard and the transfer of train work and staff from other yards was a planned affair of considerable magnitude — the whole area modernisation was probably unparalleled in scope and size. The yard, diesel maintenance and servicing depot, freight terminal, and the new entrance and exit junctions, the new Aldwarke connection between the side-by-side GC and Midland main lines, the simplification of junctions and signalling at Mexborough, and the amalgamation of the two passenger stations at Sheffield, with the closing of the GC Victoria, cost around £11m and movement savings were estimated at approximately £1.2m per annum plus savings in equipment which included 35 locomotives, 1,700 wagons, and 250 cartage vehicles. In addition many smaller yards and goods depots were closed and land made available for sale, the proceeds making a handsome contribution to the capital outlay. The yard was opened by Dr (later Lord) Beeching on 29 October 1965, but his policy of retrenchment in other directions inevitably reduced the amount of traffic throughout the system and through Tinsley Yard. In spite of efforts by diversion to bolster the throughput, the subsequent general recession, particularly in the heavy iron and steel industry, has taken its toll. The yard is working under capacity and 20 years may be its life-span in its original form.

Blast Furnace Land

In 1838 the population of Scunthorpe was no more than 303 and the essential factor in the town's phenomenal growth was the quarrying of ironstone and, from 1866, the making of iron. The iron ore field was the largest in Europe, it was easily worked, it was near the surface, fuel was easily obtainable from the South Yorkshire coalfield, the excellent ports on the Humber and Trent were relatively close, and the railway transport was good and was to be developed. By the 1930s a 'New Yard' for marshalling inwards and outwards wagons had been provided, a new passenger station built in the residential quarter named 'Scunthorpe & Frodingham' to replace the demolished original 'Frodingham & Scunthorpe' which had more than served its purpose in an out-of-place situation cheek

Rationalisation at Sheffield showing the location of Tinsley marshalling yard in relation to the former Midland and GC routes.

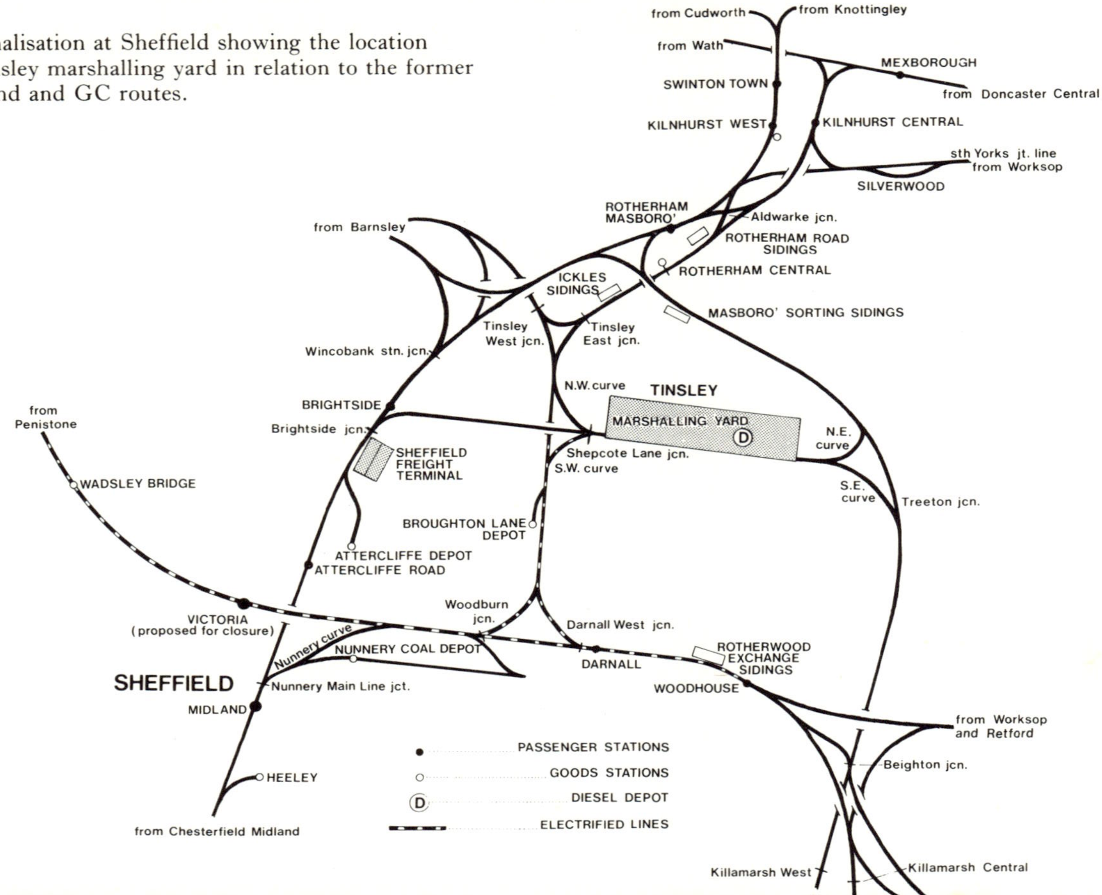

by jowl with one of the main iron and steel works. Much pioneering had been done and smaller firms had amalgamated but by the pre-war period in 1938 the four predominant firms were the Appleby-Frodingham Steel Company (part of the United Steel Group), Richard Thomas Ltd, and John Lysaght Ltd all of which became part of the British Steel Corporation in more recent times. The fourth was Thomas Firth and John Brown Ltd. According to official 1937 figures 2,760,101 tons of ironstone were mined and there were 14 iron furnaces in blast and 22 open-hearth steel furnaces which produced 1,043,200 tons of pig iron and 1,299,200 tons of steel ingots and castings. Iron ore was also brought in from the Lincolnshire and Northamptonshire beds and imported through the port of Immingham from Spain, Brazil and Scandinavia.

The iron and steel despatched by rail from Scunthorpe took a varied form and required special types of wagons to convey it, the ordinary high-side, plate and bogie-plate wagons for sheet steel, and wagons having one bolster, or two bolsters, twin bolsters, and even four and five bolsters known as 'Quads' and 'Quints' for carrying long steel – girders, bars, blooms, billets, angle tees, plates, ingots and pig iron. Industries supplied were shipping, building, railways, motor cars and engineering. Demand for one kind of wagon or another varied; there could be a run on bolster wagons followed, by a quick switch, to what were called 'highs' or what the firms called in their jargon 'box' wagons, so that a close control had to be exercised on the movement of the 12 pilot locomotives and shunters ferrying supplies into and out of the steel works.

The railway sidings at Scunthorpe consisted of a double-ended yard known as Trent, between Trent Junction and North Lincoln Junction and the 'New Yard' situated between Yard No 1 signalbox and Scunthorpe station signalbox; it consisted of sidings 1 to 12 for inwards traffic for firms, two reception roads, and sidings 15 to 23 for outwards wagons either eastwrds or westwards. Large and extensive as were the LNER sidings at Scunthorpe the accommodation belonging to the four big steel companies was equally or more so. The Appleby-Frodingham company alone had 100 miles of permanent way and employed 63 locomotives and 2,677 internal wagons, being quite a railway on its own account. Appleby-Frodingham had four entrances into its works and a fifth was constructed just before the second world war. Wagons for Appleby-Frodingham were labelled at source to entrance A, B, C, D, or E, according to the type of traffic conveyed.

The town of Scunthorpe and the works at Frodingham and Normanby Park were on a plateau running north/south, while the railway main running lines and sidings ran west/east, trains having to climb into Scunthrope from Keadby and the west, and again out of Scunthorpe to Appleby, Barnetby and the east; in both cases assistant engines were provided to bank freight trains in the rear. The movement problems at Scunthorpe were great, indeed they were exceptional. Geographical features already mentioned inhibited expansion and the place could only be kept in a state of reasonable fluidity by a relaxation of traditional methods; it is doubtful if there was or is another place in the country with so much concessional operation in which facing-road working, running round, and propelling up to 45 wagons (loaded) at a time and even in the opposite direction was carried out to such a degree; on some tracks, railway-owned and privately owned locomotives intermingled and it was freely agreed that if any form of fixed semaphore or light signalling were installed it would slow-up and 'stop the job'.

In 1938 the revenue, growth potential, and operational problems at Scunthorpe caused railway management to increase supervision at the top; the yardmaster was given three assistants instead of one, to allow round the clock coverage. An essential feature of the assistant yardmaster's job at Scunthorpe was getting round the various yards and works, all of which were rail served, keeping a close watch on the working shunters, the 12 travelling pilots which took traffic into and out of the works and transfers from yard to yard, telling them what to do next, having regard to the overall situation and the avoidance of congestion at any point. The appointment of three assistants from elsewhere instead of making local appointments by upgrading paid off handsomely.

The population of Staveley, an urban district in North Derbyshire, was about 18,000, with a very ancient church and equally old grammar school. Not as big as Scunthorpe in iron ore and steel terms it did have a commodity which was not produced at Scunthorpe and that was coal, so the combination of the two products was not inconsiderable. The main firm was the Staveley Coal & Iron Company and adjacent was the Sheepbridge Iron Company. Local collieries were Grassmoor, Bond's Main, Hartington, Renishaw, Ireland, Markham, Arkwright, Pilsley, Tibshelf, Williamthorpe, Oxcroft, Glapwell and Bolsover. They were serviced, as were the iron companies by both the LNER and the LMS companies, the concentration yards being respectively Staveley Central (GC) and Barrow Hill (Midland).

Staveley Central station, the winner of many special prizes for the best kept station competition, had four platform faces, two of them serving up and down main line trains and two serving the up and down Chesterfield branch or loop. There were four junction signalboxes – Staveley North, Staveley South, Duckmanton North and Duckmanton South, the two latter controlling the route to and from the 'LD' line running from Chesterfield Market Place to Lincoln. Markham and Ireland collieries were served by single lines terminating in the up yard at Staveley South. Staveley Central marshalling sidings or holding sidings totalled 34 in all with a total capacity of 1,279 wagons. The yard and passenger station was at the foot of a 1 in 100 climb to Pilsley, a distance of eight miles, and most through freight trains required assistance from a banking engine in the rear. With the heavier wartime traffic throughout the country and interruptions in movement arising from air-raid warnings or enemy damage there was often delay awaiting a banking engine or relief crew to take a train forward. In order to ease the pressure a suggestion was made that a proportion of the trains should be booked, or arranged on an ad hoc basis by the control office, to run over a diversionary route via Killamarsh Junction, Clowne, Langwith, the Leen Valley Extension and back to the GC main line at Kirkby South Junction north of Annesley. This was adopted and it became a timetabled feature.

Until the outbreak of war in 1939 there were a fair number of passengers using Staveley Central on either expresses or on the local Sheffield-Chesterfield-Nottingham trains, many being workmen to and from their working places or for shopping. The service was severely curtailed in 1940 and not a single passenger train ran between breakfast and tea-time until it was agreed that an up and a down Marylebone-Manchester express should make a stop. Other through expresses were the Marylebone – Bradford via Penistone and Huddersfield, Newcastle-York-Southampton, Newcastle-York-Cardiff/Swansea, York-Swindon (the erstwhile Aberdeen-Penzance), and the 10.13pm York-Bristol all of which served Sheffield Victoria, except the latter which ran via the Attercliffe-Darnall curve; any passengers from Sheffield area joined the train at Rotherham Central.

The stationmaster/yardmaster/goods agent at Staveley Central had a staff of four clerks, three inspectors, 13 signalmen, 34 guards, and others bringing up to a complement of about 80. One of the troublesome chores of the stationmaster, and for which he had neither credit nor reward, was that he was the LNER Estate Agent's representative for no fewer than 115 railway owned houses, 15 other lettings, and 107 garden and allotment plots. Changes of tenancy, and burst pipes in outside lavatories in winter-time caused a lot of work in contacting plumbers and/or in correspondence.

In the post second world war years Staveley, Barrow Hill, Sheepbridge and Chesterfield, like so many other places, felt the chill winds of economic changes in the main industries of iron and coal with some collieries closing down or, in the case of collieries being dually served as one line was eliminated. Coal output generally was less as domestic users switched to easier forms of heating. Eventually the remaining traffic on the Great Central route was switched to the Midland; the former was closed in 1963 except for a single line from Killamarsh to Duckmanton to serve the Arkwright Colliery.

Staveley Town, as the Great Central station was called when built in 1898, later renamed Staveley Central to distinguish it from the Midland station, had nearby a very good and commodious engine shed for maintaining its stud of engines. With the advent of diesel power based on the new depot at Tinsley its work was run down and final closure came with the closure of the Great Central main line and transfer of workings to the nearby Midland depot at Barrow Hill.

3
THE RAILWAYS IN WARTIME

After a year of unease after the Munich agreement of 1938 the second world war eventually became inevitable and on 1 September 1939 the Government took control of the railways; hostilities were declared formally two days later. From then on activities were immediately transferred from peace to wartime conditions and thinking. The Minister of Transport appointed a Railway Executive to be his agents and the Operating Member was V. M. Barrington-Ward (later Sir Michael), Superintendent of the Line for the LNER's Southern Area (Western Division), otherwise the Great Northern and Great Central sections.

What was in store? On 13 September an air raid on Tyneside was frustrated and three days later raiders were over Edinburgh and the Firth of Forth, dropped a bomb on HMS *Southampton* but the Forth railway bridge remained unscathed. The next morning the Germans indicated their long range ability by raiding Scapa Flow and the Orkneys. The East Coast authorities realised just how very vulnerable a place like Scunthorpe, or Tees-side, was, with a railway marshalling yard complex side by side with acres and acres of blast furnaces and steel works. They were quite unable to disguse their operations, or to put out the glare when running off the molten metal into the open pig-iron beds or when the little slag trains tipped their ladles and emptied the molten slag residue on the top of the tip to run down in a wide golden stream, lighting up the countryside for miles around.

Air Raid Precautions (ARP) became a top priority throughout the country and so far as railways were concerned the stationmaster or yardmaster was made responsible for setting up such an organisation, and the same applied to the Local

Defence Volunteers. Drilling in the early stages took place with brake-sticks and with sawn-off shunting poles until the Canadian rifles eventually arrived in 1940 to be followed by a bulk supply of denim uniforms, and the local stationmaster found himself signing military receipts for those items and for hundreds of loose buttons which had to be counted and checked like cash. Essential lighting was reduced to a minimum and turned off on receipt of an air raid alert until the 'all clear' was received. The LDV men patrolled railway lines, bridges, viaducts, and key installations to give warning in the event of enemy attack or damage and such a possibility seemed nearer when, in May 1940, Germany invaded Belgium and Holland, and France gave up the fight soon after. 335,000 allied troops were rescued from Dunkirk and dispersed, for which the railways ran no fewer than 620 special trains in 16 days getting the troops away from the stations on the south coast. Many were routed down the Great Central main line from the Southern and Great Western; they ran at close intervals, stopping en route for attention to locomotives, for coal, water, or a complete change of both locomotives and train crew. The soldiers, crowded in compartments, were at the windows, but were in no great shape, had few belongings, and welcomed with outstretched arms any gifts of food, fruit or drink which the local inhabitants could pass up to them.

It was then that the Local Defence Volunteers were renamed as the Home Guard as, following the Dunkirk evacuation, an attack by German troops on the South Coast was expected, but never came.

The railways had already arranged before war broke out for London headquarters and District headquarters' essential staff to be accommodated out-of-town into assumed safe areas; on the LNER this meant that the Train and Traffic Control and operational offices were relocated:

King's Cross District	to Knebworth
Doncaster District	to Bawtry
Lincoln District	to Blankney
Leeds District	to Driglington
Manchester District	to Godley
Nottingham District	to East Leake
H.Q. Central Control	to Gerrards Cross

Each office had been specially built of concrete with steel doors and with steel shutters over the windows. During darkness the Control office was clamped down, literally battened down; if the atmosphere became such that a 'blow-through' was necessary then the lights had to be put out and operations suspended before a door or a window could be opened for a minute or so. This went on for four years with staff working 20 days out of every 21; it was only possible to have the 21st day off by working 12-hour shifts on the seventh and fourteenth days.

They were strenuous times; it was not just a case of hours in the office for it was a case that every 60 seconds required full attention. In a Control Office, District or Central, it was impossible not to be on the ball, for incidents were taking place all the time. That sort of work is very different from that performed by many people who work, say, on a 9 to 5 basis in the same office with no outside contacts, or perhaps confined to local contacts. Certainly it could not be compared with a business executive whose Works closed at 5.0pm until the next morning, only to start up again turning out the same products. Railway freight movement was continuous and, as seen by the District Control and more so by the Central Control it was constantly changing pattern over the 24 hours; the only similarity was that the basics would be repeated during the following 24 hours but then only to a degree according to traffic volume and whether or not there had been any disturbance. At any hour of the day during those hectic days scores of freight trains were on the run but, one hour or so later, some trains on all routes would have terminated, more would have started, while those in-between would have notched up several more block signal sections.

The area supervised by the war-time Central Control at Gerrards Cross was known as the Western Division of the LNER's Southern Area and comprised the old Great Northern and Great Central lines, and some Joint Lines, starting from King's Cross and Marylebone and going northwards, embracing Peterborough, the whole of Lincolnshire up to the south bank of the Humber (plus the New Holland-Hull ferry) across to Manchester, Liverpool, Birkenhead, Connah's Quay, Wrexham, Wigan, St Helens, Southport, and the West

Riding of Yorkshire, bordered by Leeds Central, Bradford Exchange, Halifax and Keighley. From Sheffield and Rotherham the territory extended to Staveley, Chesterfield Central, Nottingham Victoria, Derby Friargate, Burton-on-Trent, Stafford Common, Leicester Central, Rugby, Woodford & Hinton (later Woodford Halse), to Banbury (exclusive). There was an operating interest in the Metropolitan & Great Central Joint from Verney Junction and Quainton Road southwards to Finchley Road at which latter place there was a freight exchange with the LMS, and also with the Great Western & Great Central Joint between Ashenden Junction, Princes Risborough, High Wycombe and Northolt Junction – south thereof the purely GC line into Neasden and Marylebone.

The staff in the Central Control was small in number, being headed by a Deputy Chief Controller (the Superintendent of the Line or with the newer title of Operating Superintendent was the nominal Chief Controller) and he, the Deputy, was the most senior man on duty throughout the whole area outside normal office hours. He was assisted by a Controller, an Assistant Controller, and a Locomotive Controller, on each nominal eight-hour shift. Accessible to each was a switchboard with 'Control only' connections, local and trunk, to York, Shenfield (for GE), and other company headquarter Controls – Derby, Crewe, Manchester (LMS), Reading (GWR) and Deepdene (Southern) the Railway Executive (Down Street) and the War Office. In addition there was a special circuit linking the Central Control with the six LNER District Controls so that train and traffic working could be supervised and co-ordinated. Central Control could ring any one District Control or have all on the line together, hearing appeals, giving decisions, and generally controlling, boosting, restricting, train, traffic and locomotive movements, bearing in mind the volume of traffic in the pipelines and the line and terminal capacity to absorb it. It was over this line that the Superintendent's morning conference was taken.

The War Office advised the Central Control of all shades of air-raid Alerts and eventually clearances affecting the whole of the railway territory supervised so there was a constantly changing country-wide picture of 'yellows' (raid imminent), 'reds' (raid in progress) and 'greens' (all clear). Most nights there was enemy action somewhere, bombs having been dropped, lines cut or temporarily blocked, necessitating diversions and this was all on top of the minor or major derailments which were natural but unwanted hazards. Apart from traffic which moved within each Divisional there were some major through flows of traffic:

Southern Railway/London Docks/King's Cross-York-West Riding-North East Scotland;
Marylebone/Southern Railway – Neasden-Woodford-Rugby-Leicester-Nottingham-Sheffield-Manchester and west;
East Anglia-north and west;
South Yorkshire/Lincolnshire and west;
Tees-Tyneside-Sheffield-Nottingham-Leicester-Banbury-West of England/South Wales/South Coast.

The Central Control exercised supervision over the express passenger services as well as freight and problems were always arising. For example a platform full of passengers including forces personnel going on or returning from leave and the train arrive 'full and standing'. What to do? A normal express into King's Cross would comprise 8, 10 or 12 coaches but war-time strengthening habitually meant loads of 18 coaches and on at least one occasion the Central Control authorised a train being made up to 21 vehicles – and the single engine still coped well! There was one case of a mixed traffic locomotive bringing 1,300 passengers into King's Cross in a train of 26 vehicles! Resource and improvisation within the bounds of safety – that was a tradition in the railway service combined with high quality steam locomotives, good steam raising coal, and a driver's skill. In war-time the steam locomotives demonstrated they could stand up to much more than the more sophisticated later electric and diesel locomotives ever could.

In late April of 1942 there was considerable secrecy and speculation regarding the despatch of a special train from London to the East Coast of Scotland; it was empty except for a number of officials and dining-car attendants who were all held in Scotland for the best part of a week waiting for a certain high personage and accompanying officials to arrive. He/they did not arrive at the time expected and the attendants had to return to King's Cross and await for a further call and which duly came; Mr Molotov, Soviet Foreign Minister arrived in Britain on 20 May and for security reasons it was considered his special train should terminate not at King's Cross, but outside London. Accordingly, Brookman's Park between Potters Bar and Hatfield was chosen. The driver of the special train drove his train very well, in fact a little too well, for he arrived a few minutes earlier than the scheduled time before the official Whitehall reception party had arrived on the station platform!

On another occasion in the early hours staff at No 10 Downing Street telephoned Central Control twice as officials wished to speak to Prime Minister Churchill who was then in his special train running down the East Coast main line. The train was stopped specially at Doncaster and arrangements made for Churchill to telephone No 10 from the Down side Inspector's office.

By December 1942 and after three years of war, and the hazards and vagaries of alerts and bombings, the railways had become very congested with additional traffic for the war effort and which brought about such drastic changes in traffic flow, changes which the railways were not originally designed and constructed to meet. The Merseyside ports of Liverpool and Birkenhead were getting snarled up with difficulties following the arrival of American personnel and equipment and administrative help was provided to the Cheshire Line commitee's offices at Liverpool by the LNER from Gerrards Cross. The military services made great demands in the movement of stores, ammunition, equipment and personnel. The many buildings and airfields constructed especially for their use required thousands of tons of basic commodities; the same materials were also required for reconstruction and repair work after bombings, or for the construction of government factories. Seven or eight train loads of bricks a day left the brickyards at Peterborough in nominated loads for East Anglia, Birmingham, Hatton, Hull, Scotland, London etc, and there was a daily train of gravel from Staines to Brackley via Northolt Junction and Princes Risborough. The USA Forces stationed near Brackley were so highly mechanised that they unloaded a whole train immediately after arrival in the short time it took for the locomotive to run light to Wooford and back to turn and take water so that the same locomotive and trainmen could take the empty wagons back to Staines for the next day's train.

Because of the vulnerability of the South and East Coast ports and sea lanes to enemy attack, much shipping was diverted to the West Coast ports which disturbed the hitherto natural traffic flows; moreover there were the difficulties inherent in ships arriving in convoy causing movement bunching after disembarkation or discharge. Big tonnages of formerly sea-borne coal from Tyne, Tees and the Humber to power stations etc, in the Thames basin and on the South Coast were diverted to rail. These trains were known as Convoy Coal No XX, most of which passed via King's Cross and the Widened Lines. In addition there were trains via Canonbury and the North London Line to gas plants and power stations in East London and to Temple Mills and Stratford locomotive depots. All this was new traffic to the East Coast main line.

The demand for steel was higher than ever before, so was the output, and this increased the railways' task tremendously. The production of iron and steel normally requires huge quantities of iron ore, pig iron, coal, coke and limestone supplemented by scrap; at this period scrap was big business as garden, house and park railings were requisitioned, torn down, cut up and sent to blast furnaces for melting down. A single furnace producing 100,000 tons of pig-iron a year required 120,000 tons of coke, 260,000 tons of ore as well as limestone and other commodities. As supplies of ore from Sweden were cut off it required the maximum development of home-produced ore, even though it had a lower iron content and produced more slag residue. All these changes taxed the railways but were partially eased by increasing track facilities at key points in the way of additional running lines and loops for temporary side-tracking. A new junction was opened at Harringay on 11 March 1940 connecting the down Great Northern line with the Tottenham & Hampstead Joint line. Another junction was put in at Sandy running north to west connecting the GN main line with the cross-country Cambridge-Bedford-Bletchley-Oxford line on 13 September 1940. A corresponding and complementary east to south junction was put down at Calvert from the same line to join up with the

Great Central's main line, thus enabling Peterborough (New England) to marshal freight trains for points on the Southern Railway such as Staines and Feltham to travel via Northolt Junction and so by-pass the normal cross-London route of King's Cross, the City Widened Lines and Holborn.

Traffic reached saturation point and movement difficulties were created not only by the additional traffic and bomb damage but also in a reduced maintenance of locomotives and rolling stock, which were being used well beyond their designed capacity. Locomotive performance in traffic was poor because of overtime 'in traffic', often standing and not turning a wheel for hours. By this time, Barrington-Ward, the Operating Member of the Railway Executive and still nominally Superintendent of the LNER Western Division, assumed overall charge of Operating and Motive Power bringing the two departments much closer together at all levels; all appointments then became joint appointments which was a great step forward. Concurrently, it was decided to set up a more sophisticated system of keeping tabs on locomotives by establishing a Locomotive Control within the District Control offices at Nottingham, Leeds, Manchester, Lincoln, Doncaster and King's Cross in collaboration with the North Eastern at York and Newcastle, and with the GE area at Cambridge, Norwich and Liverpool Street (Stratford), and later with the strengthened administration on the Cheshire Lines at Liverpool. At this time the GC/GN had 2,300 locomotives, the GE 1,335 and the North Eastern 1,788, some of which worked to the GC and GN just as, of course, the latter's engines went to the North Eastern.

The freight traffic situation on the whole national rail network in 1943 was one bordering on stagnation, congestion, constipation – slow movement because of the wartime blackout, enemy action and extra traffic. A special review was undertaken as applicable to the LNER, GN and GC area. A comparison of the four weeks ending 2 October 1943 with the corresponding period in 1942 revealed that neither weather conditions nor serious accidents had any appreciable effect on traffic working. The decline in train speed and the main cause of delay was the slow acceptance at terminal points which may have been a receiving station/depot/private siding, or a marshalling yard. The congestion in marshalling yards with the substantial increase in the 1943 traffic resulted in a slower rate of acceptance, with delays reflecting on following trains and the resort to stabling or storage of trains in an intermediate refuge which, again, affected other through trains requiring longer running margins from one refuge or bolt-hole to another to clear the line for a passenger, military special, or other still more important train.

Summarised , the stock figures for those six main routes were:

Traffic for	*Daily Average 1942*	*1943*	*% Increase*
Great Eastern Section, East Anglia, via Whitemoor	1848	2575	39.3
Great Western via Banbury	1148	1220	6.3
West of Sheffield via Woodhead	1712	1982	15.7
Peterborough north and south including King's Cross and Southern Railway	3832	5881	53.5
Neasden, Marylebone, and Southern Rly via Neasden/Feltham	471	790	67.7
North Eastern and Scotland via York	665	1545	132.3

A much larger amount of train and traffic regulation was found to be necessary in 1943 than in 1942 and consisted the imposition of 'stops' on both operating and loading. Operating Stops were designed to bring the line occupation back to normal, hopefully, within hours, but they *followed* congestion which caused the decrease in train speed. The pressure of traffic waiting to come forward caused the Operating Stop to be lifted at the earliest possible moment so there was no change of subsequent trains running at a relatively higher speed, which might have offset to some extent the loss of speed caused initially. A number of loading stops were applied, but they also followed congestion, mainly at a terminal/discharge point, and they too had to be removed at the earliest moment to reduce the disorganisation created at the production end, in manufacture and industry generally. During the four weeks of September 1943 there was an increase in military personnel and freight trains on the GC and GN:

	1943	*1942*	*% Increase*
Personnel	360	324	11.1%
Freight	290	193	50.3%
Total	650	517	

The increase in military personnel trains also caused an increase in empty coaching stock working as the requirements were unbalanced. The 50.3 per cent increase in military freight trains and which had preference over all other freight trains except No 1 fully-braked express goods/fish/meat etc., resulted in a large number of less important freights being shunted and, in the aggregate, a lower average speed for all trains. The nature of the traffic on the military freight specials – ammunition, bombs, guns, essential equipment, cement for bunkers – prevented these trains attaining any considerable speed which would have helped to offset the lower overall speed of the sidetracked trains.

The King's Cross District of the LNE Southern Area was, like the Liverpool Street district and the Southern and Great Western railways, very much in the front line during the war, not necessarily on account of bombings alone but more because of the general melee of bustle. The King's Cross operating district in 1944 comprised the East Coast main line as far as Grantham (inclusive), the intermediate branches, and the Great Central main line from Marylebone/Neasden to Culworth, a few miles south of Woodford Halse in Northamptonshire. The two main lines to the Midlands and the North had completely separate services and had no direct physical connection between them. If for any reason locomotive power or rolling stock had to transfer it was done via the Harringay war-time curve, the Tottenham & Hampstead Joint line, Willesden Junction, Acton Wells Junction (reverse) and Neasden – otherwise it was via Grantham and Nottingham Victoria! The Great Central line included the passenger and freight services which ran both via the GW & GC Joint Line between Northolt Junction on the London side of Ruislip, and

Controller's nightmare – 2. The trains which were not to be delayed. *Above* The Flying Scotsman headed by Class A3 4–6–2 No 4472 *Flying Scotsman* on the first non-stop run from King's Cross to Edinburgh approaching the Hadley Wood bottleneck on 1 May 1928. *Below* The up Silver Jubilee high speed streamliner leaves Hadley South tunnel as the two tracks become four at Greenwood. (L&GRP/David & Charles)

Above Up partly vacuum-braked express freight at Sandy hauled by Class K3 2–6–0 No 113 in the 1920s. The LNW Bletchley–Cambridge line is on the right which had crossed the GN line beyond the bridge in the distance. It was just north of Sandy where a wartime spur was laid in from north to west to link the GN to the LNW route. (L&GRP/David & Charles); *Below* The fully brake-fitted fast freights on the GN main line, able to run at up to 60 mph, were often sufficiently important to be rostered for Pacific power. The afternoon down Scottish goods from King's Cross, headed by Class A4 4–6–2 No 60025 *Falcon*, speeds down the by then widened section of the GN line between Hadley Wood and Potter's Bar on 13 June 1961. (Derek Cross)

Ashenden Junction north of Princes Risborough, and freight and LNER passenger services over the Metropolitan & GC Joint Line between Harrow-on-the Hill, Rickmansworth, Aylesbury, Quainton Road and Verney Junction (the Metropolitan looked after its own passenger trains and the freight service between Harrow, Rayners Lane and Uxbridge).

The King's Cross district also provided engines, crews and worked some 60-plus freight trains daily in each direction to the Southern Railway marshalling yards and depots, the majority passing via King's Cross Widened Lines, Farringdon Street and Holborn with a banking engine from Farringdon Street up to Snow Hill Bank.

Troop trains and stores trains passed through London but missed the centre by travelling via the North London Line and reversing at Canonbury Junction between Finsbury Park and Highbury. Ambulance trains of wounded soldiers reached the LNER via the Bletchley, Bedford, Sandy route for the injured to be detrained at Langford Siding near Biggleswade for the Three Counties Hospital. Early in 1944 there were meetings in preparation for 'Overlord' the Allied Forces projected landing in France. This invasion took place in the early hours of 6 June but the build-up was a massive country-wide operation. In the two months before 'Overlord' no fewer than 22,459 special trains ran conveying troops across the country, thousands being for 'Overlord' itself. The movement of stores and equipment for D-Day started on 10 May necessitating a further 800 special trains, and another 6,000 wagons passed on ordinary freight services. It was an unprecedented wholesale movement necessary for the construction of the airfields in East Anglia and Lincolnshire, for three-quarters of the nation's bomber air bases were located on LNER territory.

Early on 13 June 1944 one of the first flying-bombs fell on the four-track Grove Road bridge of the LNER near Stratford in East London, causing much damage. This heralded the V1 onslaught on London from the night of 15/16 June onwards, no fewer than 1,074 falling on railway property, 226 being on the LNER. Then followed the V2 rockets which came without warning and again many fell on railway property in London and South East England.

But gradually the tide turned and on 27 March 1945 the last V2 rocket fell in Southern England at Orpington and on 8 May the war in Europe ended. The railways had survived but were suffering the ravages of the previous six years. The fact that the war was over did not mean the end of the problems on the railways for it took most of the following decade to overcome the backlog of repairs and maintenance and with shortage of investment finance, the upheaval of nationalisation and the shortage, particularly, of steel deliveries for new equipment, new locomotives, and stock were slow to come and not until the massive modernisation programme of the mid 1950s did re-equipment really get under way.

4
TRAIN AND TRAFFIC CONTROL

The origin of Train and Traffic Control dates back to 1907 when Cecil Paget, the then newly appointed General Superintendent of the Midland Railway and who later was knighted, in his successful efforts to turn chaos into order in the South Yorkshire coalfield, established a small office staffed continuously at Rotherham (Masborough), telephonically connected with all signalboxes etc in that locality.

Without co-ordination there is inevitably the tendency for everybody to do his or her own thing and this is what had happened in that busy industrial area. Train and traffic control proved to be successful and the system was extended so that it ould be said that by 1912 the Midland Railway had the finest operating organisation then in existence. Since that date, this operating practice has been followed by all railways, but with variations. The old Midland Division of the London, Midland & Scottish was approximately the same size as the Western Section of the London & North Eastern covering former GN & GC lines both in mileage and density; the type of traffic carried was very similar, and so were the ton-miles, around 245 million.

'B W' – Barrington-Ward – had his operating apprenticeship at Derby under Cecil Paget, picked up a thing or two and improved upon them mainly in size of the control district and in the means of communication. Whereas the Midland operated 24 small control districts the equivalent-sized GC and GN settled for six, which naturally enabled the LNER Controls to have a much bigger picture. For example, while over a main line distance of 56 miles the Midland had Control offices at Cudworth, Masborough, Staveley, Westhouses and Toton, the King's Cross Control looked after the East Coast main line for 110 miles; Doncaster, the next Control to the north, had it for 50. This was an enormous advantage. For communications the Midland operated on many omnibus box-to-box telephone circuits which were in common use by everybody and for any purpose. But even before Barrington-Ward's day, Paget's experience at Masborough in 1907 was not lost on old Great Northern as that company followed suit in 1913 by opening Control offices at Nottingham and Leeds to co-ordinate and supervise the goods, mineral and coal traffic and trains in their respective areas. The former was located at the original Great Northern station, London Road (Low Level), and the latter at Leeds Central which was GN and L&Y Joint. Nottingham controlled the line from Grantham (exclusive) to Derby (Friargate), Burton-on-Trent, Stafford Common, the intermediate branches in the Leen and Erewash Valleys with their many collieries being serviced from the large marshalling yard and locomotive shed at Colwick. Leeds Control looked after the GN interests from Doncaster (exclusive) to Leeds, Bradford, Keighley, and Halifax via the various routes and complex branches notable for their gradients, as many as 13 being steeper than 1 in 55. Also in 1915 the Great Central opened what it called a Train Despatch office at St James' Bridge, Doncaster, but it dealt only with the cross-country lines from Penistone to Barnsley, Mexborough and Scunthorpe, and Worksop-Retford-Gainsborough-Barnetby.

It was the development of the telephone which really boosted the institution of Control offices. Like the Midland, the GN and the GC it was also taken up by the LNW, the L&Y, the GE and the NE but by no means on the same basis although they appeared to have one thing in common – they did not normally deal with passenger trains. Middlesbrough (Newport) was opened in 1910, Hull in 1923, while York Main Line Control (1922/23) would not touch anything on the York branch lines.

Practice on constituent companies was extremely fragmented until after the 1923 grouping. Early experience on the railway was in this fragmented period and one witnessed some chaotic traffic working and the setting up of District Control offices was amply justified. For years the movement of loaded and empty goods, mineral and coal trains (the term freight had not been introduced) had been governed but not controlled on the Great Northern by four strategic yardmasters who had established themselves over many years – Town at Hornsey, Cox at Peterborough, New England, Keys at Doncaster and Lee at Colwick. They were entrenched in their positions and they, or their inspectors, influenced the main streams of traffic without any co-ordination. Any traffic, or private owners' coal wagons, surplus to the booked timetabled service – and there was a good deal for obviously the timetable clerks had not got their sums right – was 'got rid of' by hooking an engine on (engines and men were plentiful) to a train load and setting it off irrespective and quite regardless of how many trains were already en route and 'on the block' awaiting acceptance by the more distant yard. For instance, the running Down Goods line from Saxondale Junction to Radcliffe was nearly always full with six trains notching up, one by one, as Colwick marshalling yard found it possible to accept and only then would Radcliffe let one out for the bottle-neck run over the Trent viaduct and get in clear at Rectory Junction.

As indicated, before the grouping the GN, GC and GE companies had made little progress in Control operation with the exception of Nottingham, Leeds, and Doncaster, and Stratford GE and Liverpool Street GE Headquarters, although both were more involved with the most intensive steam-worked passenger service in the world than with freight traffic.

In the post-grouping period, William Clow the Superintendent of the Line on the Great Central also took over the Great Northern to form the Western Section of the new LNER Southern Area, and Colonel Mauldin took over the GE for form the Eastern Section, both being answerable to the Divisional General Manager. Clow was succeeded by V.M. Barrington-Ward who had been District Superintendent at Middlesborough and he soon set to work to organise a co-ordinated Control system. District Controls were set up at Manchester (London Road, now Piccadilly) in 1924, Sheffield 1924, Cambridge 1925, King's Cross 1927, Leicester Central 1928, Lincoln 1928, Fenchurch Street 1933, Norwich 1944, and High Wycombe jointly with the Great Western on 8

March 1943. In addition the Doncaster GC office was extended to embrace the Great Northern in 1927. So by the end of the second world war the whole of the LNER Southern Area was tied up but, rather before then, there had to be a re-appraisal in the light of experience. In the case of Sheffield the area it controlled was part of the Manchester District. Likewise the area controlled by Leicester was in the Marylebone district.

In 1929 the complete re-organisation and the formation of a self-contained Control system was instituted as follows:

1. Combination of the Manchester and Sheffield Controls at Manchester.
2. Combination of the Leicester and Nottingham Controls at Nottingham Victoria.
3. Formation of a Central Control at Liverpool Street to co-ordinate the District Controls at King's Cross Nottingham, Lincoln, Doncaster, Leeds and Manchester with effect from 20 July 1931.
4. Similarly, the GE 'Headquarters Control' co-ordinated the work of the Cambridge, Norwich and Stratford Controls, subsequently adding in later years that of Fenchurch Street.

Later there were alterations following changes in Regional boundaries after nationalisation.

Barrington-Ward was a good administrator, not only in getting the best operational performance in outstanding events which took place in his time as Superintendent of the Line such as the introduction of the Coronation, Silver Jubilee, the West Riding expresses and the Queen of Scots Pullman trains and, on the freight side, the introduction of through trains of steel from Scunthorpe to South Wales in competition with a more lengthy coastwise transport, but also from a staffing angle for he was well-known for having salary economy in mind when making appointments, almost to the point of being niggardly or parsimonious. Many an appointment was made in those days when the successful applicant was given £10 a year less than the advertised minimum for the post, and given to understand he was lucky! He was never enamoured with any member of staff who, to use his own phrase, was plainly 'chasing the money'. But to be fair and with his keen enthusiasm for Train and Traffic Control he personally interviewed the short-listed applicants for the junior positions in his Central Control and made the final choice. Thereafter, he believed in giving those young men the facilities for widening their geographical and operational knowledge. For a Londoner this was important but for one who had already had country or District experience it was not so, but it was helpful and eased the transition, say, from an Assistant Controller in the Central Control to a Section Controller in a District Control Office. Leaving an office which surveyed a wide area of traffic movement – London (King's Cross/Marylebone) in the south to Humberside, West Riding, Merseyside, including the East Midlands, Derbyshire and South Yorkshire coalfields – and getting down to the detail of a coal-mining locality can be a traumatic experience such as a Section Controller on the No 3 Section in the Doncaster Control Office would have. That Section Controller covered the lines west of Wath marshalling yard, that is, from Aldam Junction through Stairfoot and Barnsley (Court House and Exchange) up to Penistone (Barnsley Junction) exclusive, together with the Stairfoot-Nostell and Stairfoot-Cudworth branches, and several collieries including Grimethorpe, Houghton, Wharncliffe Woodmoor, Monckton Main, Barnsley Main, Wombwell Main, Wentworth, Dodworth and Silkstone. The position covered the Worsborough branch from Wombwell Main Junction to West Silkstone Junction where it joined the Barnsley-Penistone line. On the whole of that No 3 Section there were no fewer than five different points where assistant engines (banking in rear) had to be provided because of steep gradients around 1 in 50, but the worst was the 1 in 40 for approximately three miles on the Worsborough branch between Wentworth Junction and West Silkstone Junction and where trains of 750/1000 tons already with two engines on, had to be assisted by the LNER Beyer-Garratt 2–8–0+0–8–2 No 2395, the most powerful locomotive in the British Isles or, if that was not available by two 2–8–0 locomotives. The very nature of the terrain worsened by colliery subsidences made it a heavy slow-moving line, with only the 5.10pm fish train from Grimsby to Ashton Moss and the meat and cattle trains from Birkenhead to York and Newcastle to enliven operations.

In contrast, the more senior No 6 Section in the Doncaster Control was, for the Section Controller, mentally electrifying. This section covered the East Coast main line from Grantham (exclusive) and the old GN&GE Joint Line from Sykes Junction north of Lincoln, the two lines joining up at Doncaster (Black Carr Junction) like an inverted wishbone. These routes had just up and down running lines with no additional goods or slow lines alongside such as existed between Peterborough and King's Cross but there were single-ended refuge sidings, some of which were converted into loops to facilitate clearing the line for the passage of through expresses. In 1949 the tons per annum per single line of track on this Control Section were:

Black Carr Junction to Newark	12,333,672
Newark to Barkstone South Junction	10,145,304
Black Carr Junction to Sykes Junction	8,255,208

which made it the heaviest worked length of route on the Great Northern and Great Central and, probably, also on the system as a whole. The task of the signalmen and the Section Controller was vastly different from that on the slow-moving No 3 Section as the very many East Coast and Yorkshire express passenger trains (35–40 in each direction daily) had to be monitored and the fast and slower moving freights interwoven, moving up block section by block section as soon as a margin between two expresses could be seen. The signalman manning the three dozen signalboxes were second to none as signalmen and telegraphists for all had to be able to understand the Morse code single needle instruments which gave them advance information on the running of expresses from 50 miles away to as close as 10 miles away which enabled them to calculate margins for slower trains to move up in front of faster trains and get clear without reacting on the latter. Indeed the busier signalboxes on the GN line might have several telegraph instruments chattering away in Morse code from which the signalmen had to pick out messages which affected them. Alertness was never more called for and particularly so in the period when the high-speed flyers were introduced and were at their zenith. The Silver Jubilee commenced in service on 30 September 1935 and leaving King's Cross at 5.30pm covered the 268 miles between King's Cross and Newcastle in four hours, an overall speed of 67.08mph with a stop at Darlington, although on its trial run three days earlier it reached 112½mph on two occasions and averaged 100mph for no less than 43 consecutive miles.

The Coronation, which was the first streamlined train from King's Cross to Scotland, started on 5 July 1937, and leaving at 4.0pm covered the 392 miles in six hours at an average throughout speed of 65.5mph, although between King's Cross and York where it made a call, the average was 71.9mph and was, by this performance, the fastest train in the British Empire. Five steamlined Pacifics liveried in Garter blue, with wheels dark red, and with lettering and mouldings in stainless steel, were built for this down and up service and one of them, No 4489 *Dominion of Canada* carried a chime-whistle and was later fitted with a bell below the chimney following Canadian practice, and given by the Canadian Pacific Railway.

There were other important named trains for the Flying Scotsman had been leaving King's Cross No 10 platform at 10am since 1862, a world record for an unchanged starting time nearly equalled by its up counterpart which had been leaving Edinburgh Waverley punctually at 10am since 1876. Incidentally, another world record was established by this train, when from 1 May 1928 it ran daily non-stop between King's Cross and Edinburgh in each direction, during the period of the summer timetable from mid-July to mid-September. Then there was the Queen of Scots Pullman which actually started as the Harrogate Pullman back in 1923, on 9 July, later being extended to Newcastle and some two years later to Edinburgh and in 1928 to Glasgow when it then became the Queen of Scots leaving King's Cross at 11.20am. The Yorkshire Pullman serving both West and East Ridings of Yorkshire was then introduced to leave King's Cross at 4.45pm and, later, on 27 September 1937 the West Riding Limited started, leaving King's Cross at 7.10pm and specially designed as a high speed service covering the 195 miles between King's Cross and Bradford in 3hr 5min – an average speed between London and Leeds of 68.4mph – notable because of the colliery subsidences north of Doncaster. There were, of course, the balancing trains in the up direction to

King's Cross, and other restaurant and sleeping car expresses both ways throughout the day or night taking top priority. Next in importance came the 10.5am meat train from Aberdeen to King's Cross Goods followed by the Aberdeen fish trains which left at 12.30pm and 1.40pm. Then there was the 6.15pm beer train from Burton-on-Trent to Newcastle (via Colwick and Newark and years later diverted to the Midland route via Derby, Chesterfield, Rotherham and Knottingley), and the famous train No 562 (later re-numbered 266) 3.35pm express goods from King's Cross to Glasgow. There were many others, all pushing their way through and overtaking the slower moving, coal, mineral, and goods trains and all had to be regulated by the controllers and signalmen.

There was great excitement on the Doncaster No 6 Control Section in the evening when the 4.0pm Coronation from King's Cross was telegraphed forward as passing Peterborough on time at 5.3½ – yes, timed in half minutes – which then meant that before it passed Grantham at 5.27½ the controller and signalmen, had to have all other trains ahead out of the way, run into loops or shunted into sidings, points reversed and 'all boards off' for the flyer without causing a check or momentary brake application. It was repeated likewise for the 4.45pm Yorkshire Pullman, the 5.30pm Silver Jubilee, and the 7.10pm West Riding Limited. The Section Controller, sitting at his desk with headphones on, had to be attentive to any voice heard on his open telephone circuit, ready to give on-the-spot decisions when appealed to by a signalman, then perhaps wondering whether the driver of a slower train being given a risky bare marginal run to a bolt-hole one block section ahead would make it before a flyer was right behind him. When the old Great Central and the old Great Northern Control offices at Doncaster were combined after the grouping it brought a fusion of staff and one member was a Deputy Chief Controller – the man in charge of a shift – a rather timid individual by nature and of Great Central extration, not hitherto accustomed to fast moving trains and more at home with the slow moving coal trains in South Yorkshire. He on tenter-hooks the whole of the time wondering whether one of the flyers would get stopped; to see his relief afterwards was a second reward to the Section Controller's own satisfaction of having taken some decision and seen it come off. Unlike the Section Controller the Deputy Chief Controller had not got his stabilisers out!

Of course, in those days, though not reaching the later wartime volume there was a great deal of traffic to move. The marshalling yards were constantly receiving traffic from works, factories, mines, steelworks etc, for shunting and despatching on freight trains at something like 20 or 30 minute intervals up and down the GN East Coast main line. At any hour of any 24 you would find that every signalman had an up and a down train in his block-signalling section, the slower coal trains constantly edging their way forward and going inside a goods line or a loop or a refuge for more important faster trains to pass.

One more word about the signalmen on the GN East Coast main line. To be a signalman at one of those signalboxes was a distinction of which they were proud and they knew that one false move, one misdemeanour in signalling, jeopardising safety, and they were OUT! Management took no risks, it was always 'safety first'. Today with centralised power signalboxes – only three signalboxes at King's Cross, Peterborough, and Doncaster control the whole of the GN main line of what was the Eastern Region before its 1966 merger with the North Eastern Region – train control is very different, for the signalmen themselves and the signalbox supervisers have a complete picture of traffic in their area. Yet even with a much reduced freight service, higher speeds of both passenger and freight trains and fewer bolt-holes mean that signalbox staff have to be on their toes if the Inter-City 125s are not to be delayed.

Locomotive Control

Those with experience of the constituents in the Grouping of railways in 1923 will be more conversant with the heritage of heavy-handed Victorian planning which, in effect, put railways and even trains and locomotives in one another's way to the point of creating utter congestion. True, there WAS competition but it was only amongst the railways themselves; speedy overnight transits were not required to the extent as in later years. Not only was there duplication of railways down the same valley, but there was an extensive use of running powers over 'foreign' lines and maintenance of locomotive sheds on foreign territory; for example, the Great Eastern and the London & North Western each maintained running sheds at Doncaster.

There was little or no train and traffic control in the accepted sense, and no locomotive control, and trains queued for hours on goods lines, one company's engines were changed over for another's at such recognised traffic junctions as Doncaster/York between the GN and NE, and Woodford/Banbury between the GC and GW. If a company did not own or lease its own shed on foreign territory it had an agreement with another company for servicing. As an example a North Staffordshire Railway engine after working a train which terminated at Nottingham Midland, ran light through the Sneinton Junction exchange, and over GN metals to Colwick GN shed, where it remained all day until the crew came out of their lodgings in the evening to work a train from Nottingham Midland back to North Staffordshire. It would have been much more convenient, and certainly more economic for the engine to have been stabled in Nottingham Midland shed but its preference for Colwick was that at some time back in history the North Staffordshire Railway was more friendly towards the Great Northern than the Midland.

This type of operation, and the inescapable fact that steam locomotives had to be constantly serviced with coal and water, resulted in movements in and out of locomotive running sheds on a colossal scale, and this can be borne out by a reference to an old working-timetable where it was the practice to tabulate booked departures of engines from the shed. In those days engines and men were available quite freely and it seems astonishing that when private companies sought to pay dividends there should have been so much wastage. The hard-headed Victorian and Edwardian businessmen must indeed have been very hard-headed because the apparent rule-of-thumb operation reigned and the science of even simple economics was unknown or unpractised. At that time it seems, it had not been driven home that wheels on vehicles were for movement – not only that they should turn, but turn as often or as continuously as possible. It was so with wagons and it was so with locomotives; there was a gross wastage of capital invested in steam motive power. This must not be taken as indicating that bad engines were built, although of course, some were. Individual mechanical engineers had their theories and put them into practice, and in this evolutionary progress British locomotives attained the highest reputation for efficiency, and maintenance was also at a high standard. Moreover they were driven by men of unequalled skill and judgement, and whose record for unobtrusive efficient and safe working has always been a source of pride to railwaymen and public alike. Technical locomotive practice was high, but operationally, it was at a low level. Briefly, locomotives stood about on trains too much and they went to shed too often.

The 1921 Act under which the railways were amalgamated into groups reduced over a hundred separate railways companies to four (with a few exceptions) and brought a train of economies; in the early 1930s came 'pooling' – an inter-railway agreed concentration at terminals with a canalising of traffic flows to secure bigger loads on individual trains. This policy produced transit and other benefits and also let the light of day into locomotive operation, mainly in the elimination of duplicate sheds, maintenance, and administration. But it was really only with the wider introduction of train and traffic control that chaos was sorted out, and wheels began to turn more freely.

Like other things, animate and inanimate, a locomotive has a life of 24 hours to each day. As with an individual so with the article, the test comes as to what is done with the time. Some philosopher (or more likely a businessman!) said 'time is money'. In the last analysis everything comes back to time, that merciless feature of the universe, from which there is no escape. So, with the locomotive, it is a case of work-work-work for as many hours of the 24 as possible.

As each of the LNER control offices opened up for train and traffic control there was a man on each shift supplied by and on the pay-roll of the district locomotive superintendent in charge of the running sheds. He had more of a watching brief than an active role and he generally supplied the technical advice in case of engine failure or assisted the local running shed foreman in arranging relief for crews. The LMS began what may be called the first Locomotive Control by instituting in 1938 the control of unbalanced engines at its Divisional Control Offices at Derby, Crewe, and Manchester. Unbalanced in this sense means a locomotive which has powered a train from A to X but there is a shortage of traffic to maintain the service from X back to A, so there is no return working. So the underlying principle of this Control was to get the unbalanced engine power at X back to its home base by picking up a load from W or Y to A, or alternatively to return light. Another important principle was that no engines other than those belonging to a home depot should be employed on that depot's diagrams, except by arrangement with Divisional Control.

The LNER in the early days of the second world war realised that much more had to be done to employ steam engines more effectively than a somewhat perfunctory effort to get an unbalanced engine home. An organisation was set up in the LNER Southern Area covering the GC and GN in which the author was deputed during the war years to devise and recommend a system for use in the area comprising the King's Cross, Nottingham, Manchester, Lincoln, Leeds and Doncaster Control districts, which was done without any increase in manpower to operate it. It consisted of a system of census-taking at a week-end zero hour when there was a minimum of movement on trains and most engines were on shed. This was followed by a continuous record of the movement of locomotives between one District Control and another and an agreed figure several times a day which gave a plus or minus balance of so many engines, from which adjustments were made having regard to traffic dispositions. Individual locomotives were not controlled step by step, as it were, as the wartime movement of traffic and the large number of engines involved made that utterly impossible. This worked very well and improvements came about after the war when traffic patterns had settled, punctuality was better and engine diagrams had become more realistic and workable – and when steam was well on its way out.

Before the second world war, the traditional organisation was on departmental lines and provded for the locomotive department being entirely responsible for the building or purchasing of engines, for the maintenance of an adequate number of engines to do the work required by the traffic department, for keeping the engines in a good state of repair, and for the proper distribution or allocation to the different depots in accordance with their needs. Under this organisation the whole of the drivers and firemen were attached to the locomotive department and under the supervision of the chief, known as the locomotive running superintendent, or mechanical engineer. More than this, the locomotive department produced its own diagrams and men's rosters, which, as expected, provided a good deal of split crewing, ie, enginemen from one depot and the guard from another. Barrington-Ward, for one, saw the folly of all this, and in his part of the country brought the locomotive running department and the traffic or operating department under him at the top and with posts below him in London and the provinces being jointly appointed.

This ripped down the iron curtain and as time went by enabled the diagramming of engines and enginemen, hitherto done by the locomotive department, and the diagramming of guards, hitherto done by the district superintendents or traffic offices, to be brought together under a central timing and diagramming office. At shed/depot level, at district level and again at headquarters level the strong feeling of departmentalism died out and was replaced with a joint common-working approach with allegiance in the districts to the traffic manager.

It must be said at the outset that Locomotive Control really starts in the diagramming section of the central timing and diagramming office, after an appraisal has been made of the booked train service, and the number of types of locomotives required to move the traffic, bearing in mind the specific, or peculiar, needs of one locality to another or, indeed, finding work for a particular class of engine. Therefore, the basis of Locomotive Control is the depot allocation list, by type and individual numbers.

The diagramming of engines on an out-and-home basis, that is, a train working comprising a run of X miles there and back which can be accomplished within the trainmen's turn of duty, had the material advantage that each engine was kept in repair by the same shed staff, and there was probably a further advantage in the far-back steam days that some drivers stuck to one particular engine, but let us admit that it was before the 8 hour day or 48 hour week was introduced. Progressively, since then, with steam being replaced by more expensive electric and diesel machines, the tendency, dictated by economic needs, has been to employ locomotives more intensively.

When the LNER was formed in 1923 the operating stock of constituent steam locomotives was:

Great Northern	1359
Great Central	1358
Great Eastern	1336
North Eastern	2156
North British/GNS	1196
Others	7
Total	7412

No direct and sensible comparison can be made with a region's operating stock at any later date because of wholesale changes in boundaries lines and means of traction, there being a steady transition away from steam from 1953 onwards. It is significant to quote the all-line stocks over the years which adequately illustrate this trend:

	1937	*1948*	*1958*	*1962*	*1967*	*1968*	*1969*	*1973*	*1980*
Diesel	35	69	1201	3683	4742	4373	4183	3633	3078
Electric	–	17	72	178	341	329	328	317	301
Steam	19679	20211	16103	8764	362	*	– *	– *	– *
Sub-total (a)	19714	20297	17376	12625	5445	4702	4511	3950	3379
MU Electric	–	4235	5270	6982	7492	7348	7332	6747	7558
MU Diesel & Diesel-Electric	–	40	2422	4087	3988	3810	3649	3467	3257
Sub-total (b)	–	4275	7692	11069	11480	11158	10981	10214	10815
Total (a+b)	19714	24574	25068	23694	16925	15863	15492	14164	14194

*Plus three narrow gauge steam.

The difference of 8,709 in the 1948 all-line locomotive and multiple-unit stock of 24,572 and the 1968 total of 15,863 speaks volumes for improved means of energy creation, and the utilisation of it in various forms, (think of the light

Above Dukeries Junction, with the GN main line at the lower level and the LDEC line on top. A spur linked the two lines on the far side of the station. (L&GRP/David & Charles); *Below* An Eastern Region coal train bound for London Transport stations pulls into the loop at Rickmansworth in February 1952. (C. R. L. Coles)

Above Diesel-hauled freight on the GN main line at Huntingdon on 1 May 1975. (G. T. Heavyside); *Below* To provide adequate braking power for unfitted freight trains BR coupled brake tenders to certain diesel classes in the 1960s and 1970s until vacuum, and later air, brakes were adopted more widely on freight trains. The brake tender is seen behind the locomotive of this train on the East Coast main line north of Durham in 1965. (I. S. Carr)

locomotive mileage saved by using MU stock!) to say nothing of the loss of traffic to road transport.

As multiple-unit stock cannot by the nature of requirements work continuously, and its use is localised, no control in the train and traffic control sense is necessary. There are enforced slack periods in between morning and evening suburban peaks, and after midnight, which enables an effective control to be maintained but the great reduction in the total stud of locomotives, from 20,297 in 1948 to 4,511 in 1969 enabled a re-appraisal to be made of the ability to exercise a tighter control, if not an individual control of each locomotive. Individual control on the Great Northern Line was started with full dieselisation in 1962 covering the King's Cross, Lincoln, Doncaster, Sheffield and Rotherham districts, and 18 April 1966 was 'D-Day' for the institution and extension of the system by embracing the whole of the then Eastern Region, – the GC, GN, GE, LT&S, Midland, – the parts retained or inherited, and the control covered the whole 24 hours whether in or out of shed, whether working or being maintained. A booklet setting out the objectives and methods was issued with clear lines of responsibility and authority defined. The records was basically a straight-line graph with the locomotive numbers down the left-hand side and a time-scale midnight to midnight along the top, thus showing in different colours how a locomotive was being occupied at any one time:

Out of depot	On train work	In depot	On repair
	Light running		Depot maintenance
	Available		Available
	Out of Control Division		In Main Works

Appropriate information and assessments were made by each level of management: depot supervisors, divisional district controllers, and regional controllers so that representations and/or decisions could be done.

Traditionally, timing of trains and diagramming of locomotives and crews has been an aspect of railway train planning that has been in support, and not necessarily in the van. Throughout the years staff engaged on this type of work have been conditioned to regarding themselves as the servants, not in any derogatory sense, of the train and traffic planners, meeting the traders' requirements of moving traffic from A to B with the minimum cost, which is after all the primary purpose of the railway. Nevertheless, though in support, they were part of the freight and passenger movement teams with a much closer association between plan and practice.

When the Eastern Region introduced 'Advanced Traffic Information' (ATI) it was the intention, step by step, to incorporate a more sophisticated form of locomotive control but the whole project was taken over by BR headquarters and included in the all-line development of 'TOPS'. (See Chapter 7).

Wagon Control and Distribution

Before the 1923 grouping each company, large or small, had its own way of regulating the use of its wagons and passenger stock. There was a certain amount of common user of wagons by arrangements through the then Railway Clearing House based on its Euston, Eversholt Street, headquarters. The Great Northern exercised a form of control from an office at Grantham; it was a loose form of control conducted by telegram or by 'UTM' - Urgent Train Message, a written note sent by passenger train.

Following the establishment of the LNER a Central Wagon Control was set up at York under C. M. Jenkin Jones whose new type circulars containing advice, instructions, pressures, cajoling were variously treated by depots with respect or derision, and variously placed on the spiked file, clipboard, fireplace, or wastepaper basket. But the name Jenkin Jones and what he stood for became familiar – almost a byword – at every level thoughout the length and breadth of the LNER system. It was an organisation that was inevitable, a 'must', but beneficial though it was, it had its faults. Each District in each Area of the LNER had its local Wagon Control but too often Central Control was only tuned in to traffic working, and to what there was of Train and Traffic Control. It could be too remote. Secondly, it had only paper teeth. Thirdly, any expected surplus of wagons (in compiling the station's daily stock report) was based on an estimate of how many wagons would be unloaded and became available during the day; the estimate was a 'guesstimate'. This so often produced an inflated surplus figure and, consequently, an inflated distribution order which could not be met in practice or, in turn, was completely out-of-step with what the booked, or even a stepped up, service of trains could convey over a given route. The next morning there would be an inquest from 'above' but, some people were so wooden or obstinate that they could not see it was impossible to move 'X' wagons if the full complement of such wagons existed only on paper.

However, the Central Wagon Control did perform a valuable service in highlighting the capital value of the wagon stock and the need to keep the wheels turning with revenue-earning traffic. It was the best that could be achieved with the facilities then available – no mechanisation, no computer, and in many cases quite inadequate telephonic communications, for most of the transmissions were over the telegraph network. In course of time it was improved and is now mainly telephonic of a high standard with direct dialling, or by data transmission with computerisation.

After nationalisation in 1948 it was a welcome and long overdue decision on the part of management to split up the Central Wagon Control at York and let each Region the Eastern, North Eastern, and Scottish, control its own wagon movement and liaise with BR headquarters at Marylebone on inter-Regional movements. The Great Northern Line took over its Wagon Control responsibilities on 20 May 1959, the new Freight Rolling Stock section coming under myself as Freight Assistant, thus providing a co-ordinated cohesive authoritative factor between Wagon Control, Train and Traffic Control, and the Freight Movement Sections. This did actually happen and for the first time the Wagon Control had teeth. Traditional empty wagon flows between Regions, tolerated by the erstwhile York organisation were adjusted and much cross-haulage and empty wagon mileage eliminated.

Reference has already been made to the heavy quantity of coal traffic moved by the Great Northern and Great Central, and the LNER from collieries in the East Midlands, Derbyshire and Yorkshire which until the outbreak of the second world war in 1939 was predominantly conveyed in wagons belonging to a host of colliery owners, large and small time coal factors, merchants, and even wagon financiers. After discharge individual wagons were either labelled, or not plated, 'To be returned empty to . . . colliery' which meant a great deal of sorting and shunting had to be done in large fans of sidings in main marshalling yards. Coal conveyed in a private owner's wagon enjoyed a cheaper conveyance rate than if conveyed in a railway owned wagon. It was, indeed, it *must*, have been a questionable practice having regarding to the additional work which the railway had to undertake in sorting, in moving and shunting wagon repairs sidings for the many firms of repairers, in wagon inspection/examination, and in documentation.

It was a welcome relief when, in 1939, as a wartime emergency, all privately-owned wagons were taken over by the Government and from which time they became common user for loading from anywhere and with anything, coal or other suitable commodity. As the years passed many wagons were life-expired and broken up, especially those fitted with old style grease axleboxes. This phasing out process took a long time. Being common user, fewer wagons were required to convey coal, more wagons with a higher carrying capacity were being built, so that thousands of former private owner wagons had to be stored in yards and on little-used branch lines waiting to be called forward by the mechanical engineer's department for breaking up. The movement of empty coal wagons back to collieries took on a fairly regular pattern, subject only to underground mining problems or labour troubles. In winter weather conditions movement had to be closely watched by Wagon Control and by Train/Traffic Control in conjunction with the Area and Regional staff of the National Coal Board. As an example of the distribution of empty coal wagons back to collieries, the table shows the pattern of booked services from the London Midland Region's stations on Merseyside via the Woodhead tunnel during the winter of 1960/61.

After the introduction of the prototype Merry-go-Round power station at West Burton in 1965, followed by others, the pattern of coal distribution and consequently the return of empty wagons was radically changed and simplified as described in a later chapter.

Payment Services via Woodhead ex LMR to ER—WTT Winter 1960–61.

No.	*Class*	*Days*	*Time*	*From*	*To*	*Day pattern M*	*T*	*W*	*Th*	*F*	*S*
267	F	MX	9.30am	Dee Marsh	Wath*		×	×	×	×	×
263	H	SO	3.20pm	Glazebrook	Wath						×
281	H	EWD	5.01pm	Heaton Mersey	Warsop SX	×	×	×	×	×	×
					Barrow Hill SO						
291	H	EWD	6.30pm	Dewsnap	Barnsley Jn	×	×	×	×	×	×
293	H	EWD	5.45pm	Heaton Mersey	Seymour Jn SX	×	×	×	×	×	×
					Warsop Jn SO						
305	H	SX	4.05pm	Huskisson	Spink Hill	×	×	×	×	×	
669	F	SO	7.00pm	Latchford O.L.	Mansfield CS						×
671	F	SO	11.15pm	Ashton Moss	Mansfield CS						×
099	J	MX	1.45am	Heaton Mersey	Barnsley Main		×	×	×	×	×
109	H	MX	12.45am	Glazebrook	Wath		×	×	×	×	×
317	F	MX	1.24am	Partington	Seymour		×	×	×	×	×
350	J	MX	12.40am	Brunswick	Wombwell Ex		×	×	×	×	×
149	H	MO	5.10am	Dewsnap	Barnsley Jn	×					
661	F	SO	2.50am	Garston	Wath Yard						×
155	F	MX	4.00am	Glazebrook	Warsop Jn		×	×	×	×	×
155	F	MO	6.25am	Godley	Spink Hill	×					
003	H	MX	4.25am	Glazebrook	Wath Yard		×	×	×	×	×
625	C	SO	4.32am	Northwich	Mansfield CS						×
163	J	MO	6.15am	Glossop	Barnsley Jn	×	×	×	×	×	×
		MX	6.12am								
167	H	MX	5.40am	Heaton Mersey	Warsop Jn		×	×	×	×	×
625	C	MO	5.30am	Northwich	Mansfield CS	×					
175	F	MX	7.10am	Ashburys	Brookhouse		×	×	×	×	×
179	J	EWD	7.48am	Dewsnap	Barnsely Main	×	×	×	×	×	×
181	J	EWD	7.42am	Ardwick	Seymour Jn	×	×	×	×	×	×
663	F	SO	6.10am	Northwich	Warsop Jn						×
191	J	EWD	9.50am	Godley	Wombwell Ex	×	×	×	×	×	×
645	J	MX	4.02am	Bidston	Cadeby Colly*		×	×	×	×	×
201	H	SO	9.05am	Cornbrook	Wath Yard						×
201	H	SX	9.15am	Cornbrook	Wombwell Ex	×	×	×	×	×	
211	J	Q	11.35am	Godley	Rotherham Road	Q	Q	Q	Q	Q	Q
589	J	EWD	11.44am	Dewsnap	Mansfield CS	×	×	×	×	×	×
721	H	MSO	11.30am	Dewsnap	Wath Yard	×					×
229	J	SX	10.40am	Glazebrook	Warsop Jn	×	×	×	×	×	
221	H	MThX	12.10pm	Ashburys	Wombwell		×	×		×	×
247	J	SO	10.00am	Walton	Spink Hill						×
221	H	MthO	12.50pm	Ashburys	Wombwell	×			×		
231	H	SX	12.14pm	Trafford Park	Wombwell	×	×	×	×	×	×
213	H	SX	1.40pm	Dewsnap	Barnsley Jn	×	×	×	×	×	×
243	H	SX	12.30pm	Portwood	Langwith	×	×	×	×	×	
245	H	EWD	2.54pm	Godley	Barnsley Jn	×	×	×	×	×	×
247	H	MWFO	10.00am	Walton	Spink Hill	×		×		×	
665	F	SO	11.59am	Mold Jn	Warsop Jn						×
253	J	SX	3.05pm	Dewsnap	Barrow Hill	×	×	×	×	×	
261	H	EWD	2.25pm	Heaton Mersey	Warsop Jn	×	×	×	×	×	×
623	C	EWD	1.50pm	Northwich	Mansfield CS MFSO	×	×	×	×	×	×
					Seymour TWThO						
263	H	SX	2.15pm	Glazebrook	Wath	×	×	×	×	×	
231	H	SO	2.10pm	Trafford Park	Wombwell						×
667	F	SO	3.56pm	Ashton Moss	Warsop Jn						×
265	H	SX	3.55pm	Ardwick	Warsop Jn	×	×	×	×	×	
						26	31	32	31	32	34
					Totals	+ 1Q	1Q	1Q	1Q	1Q	1Q

24½ ton John Summers Wagons
Runs if required

WTT 6267 for Yorkshire Main

5
RATIONALISATION

Speeding the Services

Early in 1947 the LNER reviewed the freight service on the former Great Central main line, mainly between Annesley and Woodford, a service which had literally gone to pieces because of war-time priorities and vicissitudes, and the blocking back by the Great Western at Banbury arising from the Great Western's internal movement problems. This had led to trains standing awaiting a clear line on the GC and to the unwanted development of crew relief at Leicester Central. It was something forced upon the LNER and it had grown just like Topsy. The aim was to speed up point-to-point times and overall running times so that Annesley engines and crews would work out to Woodford and home within a shift. Before the war crewing from Staveley and Annesley was on a lodging basis, men going to lodge in dormitories for 8, 10 or 12 hours before returning home; it was a system which broke down during the war on three counts – delays en route, trainmen did not relish being away from home too long when air-raids were always possible, and food rationining. Before long Annesley men never got beyond Leicester to the south and Woodford men never got beyond Leicester to the north. Overall speed was reduced, paths were lost, trains were cancelled, engine availability was reduced by longer hours in traffic, and more train crews in all were needed.

Train crews accepted the ideas and were enthusiastic over the new scheme of through working which started on 30 June 1947. As a non-monetary bonus, Management agreed that the men could sign off at Bulwell Common if under the 8-hour day as practically all Annesley crews lived at Bulwell and Bestwood. Except for the retention of certain express freight trains the entire freight service was re-cast to provide for 31 trains daily in each direction, all running at a uniform speed. Maximum loads were 50 loaded mineral, or 55 loaded goods, or 60 empties, or equivalents, hauled by 2–8–0 locomotives.

Just over 10 years later the GC main line, as a penetrating line, was transferred from ER to LMR operational control on 1 February 1958. The Eastern Region in its locomotive building requirements included 45 new BR standard Class 9 2–10–0 locomotives for the Annesley-Woodford service and they became available in December 1957 to replace the 2–8–0s. In spite of a strong temptation to hang on to them the Eastern did the honourable thing by putting them into the service on 30 December and so they passed to the London Midland Region a month later.

It was appreciated by all that a large measure of punctuality was necessary and, because of some back-sliding after the service started, instructions were given that if any train engine for Woodford whistled up late at the locomotive depot outlet signal – even one minute late – the signalman at Annesley No 4 Box was not to let the engine off the depot. For a day or two it meant several trains were cancelled but these were still fewer than under the old working. After a few protests the Annesley locomen saw the sense of it and as they did not want to lose a run and be put on to some dirty shed duties they made a point of seeing, with the fireman, that their engines were in full steam, properly tooled up, and whistling up at the outlet signal before or on time. We rigidly maintained the principle of 'right time or not at all' and the result was so impressive, not only on the regular right time starts on the complete Annesley-Woodford service, that it also had a marvellous phychological effect on the crews of engines to work trains to other destinations. It was truly remarkable. To see departure record sheet at Annesley Yard was like looking at a passenger train departure sheet – a long repetition of 'T' for right-time away.

Brian Perren in his *British Express Freight Trains* said 'This Annesley-Woodford service, unique on British Railways, undoubtedly is the peak of unbraked steam operation in this country', and Colin Walker in his most excellently produced *Main Line Lament* wrote 'It was a step towards greater efficiency and its success was dramatic' and '. . . as a freight service it was incomparable and became a showpiece of efficiency'. The principle of right time or not at all was introduced consequently at a number of other locomotive depots and yards and was equally successful.

The economies in various ways were very considerable. Staff establishment was cut by 37 sets of enginemen and 25 guards totalling 99 men. Train miles per train hour went up by 100%. The Leicester Central relief scheme was the first to be abandoned and the aim then on was to eliminate all the others.

In 1949 the total locomotive operating stock of all classes was ER Western Division (GC and GN) 2,300, Eastern Division (GE) 1,335, making an Eastern Region total of 3,635. Although four years had passed since hostilities ceased there was a heavy backlog of repairs because of a shortage of material and new equipment for locomotives and rolling stock generally. It was seemingly impossible for the chief mechanical engineer's department to get on top, so there was a constant drive in this direction as well as in train running performance, both passenger and freight. On the freight side much was being done to curtail weekend working which had been a feature from the beginning of the war to speed individual trains or a whole block of trains between two points by breaking down the intermediate crew relief schemes which had been evolved during the war years at Leicester Central, Bulwell Common, Staveley Central, Sheffield (Darnall), Penistone (Barnsley Junction), Mexborough, Lincoln, Newark, and Hitchin.

Before 1935 all points east of the Pennines despatching trains via Woodhead tunnel had to marshal their trains in sections before leaving for Merseyside, to detach at Godley or Guide Bridge, or elsewhere. To overcome this a new yard was opened between Dinting and Mottram on 28 October 1935 to receive all non-through trains rough as they came and, while being shunted by gravity, the locomotive went round to the west end of the yard and took out a formed train destined for one of the many terminals surrounding Merseyside, the Wirral, and Lancashire areas. The new Mottram Yard, 11 miles from Manchester, had eight reception lines, each capable of holding 80 wagons, on a natural gradient of 1 in 85, followed by 20 sidings each holding 65 wagons, with points electrically controlled and operated by pneumatic power. As the main running lines were on a down gradient the yard alongside followed suit so with gravity shunting and no pilot shunting engines it was possibly the cheapest run marshalling yard in the country, in 1961 it dealt with 74 trains daily. It was strange that until Mottram was put on the marshalling yard map by Barrington-Ward, there was not a marshalling yard in the accepted sense dealing with traffic in that direction, but only the odd sidings at Godley and Guide Bridge, and they acted more as lay-bys. The reverse direction was well catered for both at Guide Bridge (Dewsnap) dead-end yard and in the through yard at Dunford. Dunford was 22 miles from Manchester and was immediately east of Woodhead tunnel and it performed the same sort of work as Mottram but in reverse; it accepted rough trains from Mersyside, shunted them, and the train engine took out a forward load. Yet it was a yard that could and should be closed and its functions be spread over other yards, both at the despatching and receiving ends. Additionally and most importantly, it would take out some 90 minutes of the journey time of every train calling there, improve the proportion of men working out-and-home between Manchester depots and Barnsley Junction and so give greater fluidity.

The proposal to close Dunford was met with opposition but the Cheshire Lines Committee and the London Midland at Crewe rowed in to help, making more through loads to by-pass Dunford and leaving the residue for Guide Bridge to deal with. The yard was closed on 8 August 1949 with net annual savings of £21,297 and capital savings on the Manchester-Sheffield-Wath electrification account of £19,370 plus savings arising out of the quickened movement leading to the abolition of the Barnsley Junction relief scheme and which could not be quantified. These sums seem small in the context of current inflation but it was not chicken-feed in 1949. About 100 freight trains a day ran in the down direction towards Merseyside with slightly fewer on the up and the proportion of Gorton locomen and Guide Bridge guards working both ways rose sharply. Meanwhile work had started on the conversion of this route from steam to electric working.

The Line Traffic Management organisation set up in 1957 got bedded down very early and good results started to show up in many directions. The success of the Annesley-Woodford out-and-home scheme then 10 years old sparked off similar schemes on other routes since trainmen refused to lodge away from home in hostels, barracks, or private lodgings as they had been accustomed to do before the war for the princely sum of one shilling and sixpence (7½p) a night. Yet in those pre-war days with the recompense so outrageously low, it was not, strangely enough, unpopular with some men. One Colwick freight guard was reputed to have 'a fancy bit' in Manchester. In 1940 one of the Staveley Central guards who actually lived at Eckington several miles away and with a passenger service not always convenient for his very odd and varied booking-on times, would gladly take over any lodging turn not wanted by one of his mates, his philosophy being that in a week's lodging turn he only had to travel from home to work and back three times instead of six as Monday, Wednesday and Friday he would be sleeping away. (Travelling to and from work at any hour of the day or night would not be very troublesome today as almost every workman has a car at his disposal; in the old days miners went to work by train and alighted at specially erected colliery platforms, but now they have car parks).

In 1933, in the author's days at Bradford, there were amongst the goods guards two who were brothers. One was the local union branch secretary who worked to the letter of the law and the other was the most co-operative soul you could find and who was somewhat contemptuous of his brother. Throughout the summer holiday months and particularly at 'Bowling Tide', the Bradford holiday week, the goods guards had to help out to work the many extra holiday passenger trains to Scarborough, Bridlington, Yarmouth, etc, and it would be necessary to 'pull them up' on Thursdays and Fridays, that is, to give them train work earlier than that shown on the weekly roster so that on the Saturday they would be available to book on duty still earlier to cover the excursion trains. This entailed having shorter rests which was no problem to the amenable brother as he would 'short-rest', go anywhere, lodge anywhere, . . . 'Just tell me where you want me to go' he would say. And if he did not 'know the road' officially he had sufficient common sense to know what to do and he followed the driver who did 'know the road'. He was one of the old school of railwaymen, like father like son. There was something inborn in them which gave them a pride in their job; they were pleased to think they were needed at all or any hour, that they were indispensable, quite unlike the 'unsocial hours' conscious men of today. It was his brother, the union man, who was the exception. This pride in their job, this happy attachment to the railway had a reward of the basic wage of fifty-five shillings per week (£2.75) but non-one seemd particularly disgruntled as consumer prices were stable.

But in the late 1950s out-and-home schemes were extended over other routes aided by more and more diesels replacing steam locomotives. An inevitable and unavoidable time-waster in steam days was at the turn-round point occasioned by the locomotive having to take water, maybe coal as well, and go on a turntable or triangle to turn, whereas the diesel was always ready to back on to a return train. Speeding up the point-to-point journey time and the turn-round time enabled a crew to work out and home within eight hours or within a negotiable period. Some of these schemes were easy to introduce but others were difficult, much depending upon the outlook of a union branch or branches of the two unions, NUR and ASLEF. The author took the chair at many of these staff consultations and recalls that in one case he had to call three different meetings at Lincoln with the best part of 50 men present, all representatives from affected depots, before the scheme was finally accepted, not because there was anything wrong with it or its presentation, but because one depot's representatives did not want to lose out to another's. In this respect one could feel an animosity or jealousy across the body of the hall as one or other representative got up to have 'his legitimate say' or to interrupt another, and yet they would persist in referring to each other as 'brothers' . . . or 'our brothers from . . .' a term which seemed hypocritical.

By April 1963 the Great Northern Line had an allocation of 287 diesel locomotives based on four depots – Finsbury Park (London), Darnall (Sheffield), Wath and Lincoln, and included 122 Type 2 (1,365hp), 36 Type 3 (1,750hp), 46 Type 4 (2,750hp), and 8 Type 5 (3,300hp). This new breed enabled tighter diagramming and quicker turnrounds at intermediate points to be achieved. It created an incentive for train crews linked with mileage payments and improved punctuality which, in turn, ensured locomotive diagrams being adhered to and regular maintenance being carried out. The higher speeds made possible by diesel operation enabled higher mileages to be obtained per turn of duty and so facilitated the abolition of intermediate relief schemes.

Having regard to all the normal principles of freight operation – the maximum user of a much more expensive unit of locomotion, maximum train loading, adequate braking power on the trains to supplement that of the diesel (the degree of braking required is greater because diesel speeds are higher than with steam operation), and meeting commercial requirements – it became clear that the existing pattern of freight services which catered for seven different train classes based on speed/braking characteristics had to be drastically revised. Too many different speeds meant higher class trains had preference at the expense of lower classes and so inhibited out-and-home crewing for many of the latter. Dynomometer car trials were held and a number of conclusions surfaced, the principal one being that the hauling capacity of all diesel locomotives was found to be in excess of the braking capacity when running in the 35–40mph range. It must be remembered that in the mid 1960s many freight trains were only partially power braked – or were not braked at all except on the locomotive at the front and guard's van at the rear. Should we reduce the load or reduce the speed? Neither solution was acceptable and it was decided that the diesel locomotive brake power had to be supplemented by traffic wagons fitted with vacuum brakes, or by using a brake tender – a specially weighted vehicle virtually the equivalent of the tender on a steam engine – which, of course, increased the dead weight to be hauled.

Trials took place and it was decided to simplify the working pattern on the GN line by reducing the classes of train from seven to two, Class 4 and Class 7. The Class 4 covered the express merchandise and block trains of fish, meat, fruit, etc, where all wagons in a train were vacuum braked. Where the 'seven star' class train (normally unbraked) required braking additional to that of the locomotive, brake tenders were used extensively but the real solution was the wider use of vacuum fitted merchandise wagons; for coal trains from colliery concentration yards provision was made for a fleet of 2,500 'Minfits' (vacuum fitted mineral wagons) to work in a pool for trains from the Mansfield area to Immingham, Whitemoor for East Anglia, and Ferme Park for Southern England. In this the National Coal Board was most co-operative. The first 'seven star' diesel operated trains were run from Doncaster to Peterborough (New England) in 1962 and saved 8,000 train miles a week, from New England to Ferme Park by April 1963 with Type 4 locomotives conveying loads of up to 1,540 tons thus reducing the original 16 steam-hauled services to nine with higher speeds facilitating out-and-home working. Other main routes for 'seven star' development were Wath-Scunthorpe, Highdyke-Barnetby, Scunthorpe-Birmingham (Norton Junction), Scunthorpe-Severn Tunnel Junction, Colwick-New England and Worksop-New England.

The reduction in freight classes from seven to two was an unqualified success, increasing the general level of speed, more out-and-home working, and with a very considerable saving in crews at the erstwhile intermediate relief points. But by 1982 most of the traffic of yesteryear had gone.

Cutting the Marshalling Yards

The intensification of attention to the improved running of freight trains also carried with it the need to watch every other associated aspect. The squad of traffic analysis investigators sent round strategic marshalling yards for a three-day appraisal came back with plenty of evidence which had to be followed through to a logical and satisfactory conclusion – improved transit, and more economic working. The first taste of postwar rationalisation was the closure of Finchley Road exchange sidings for freight traffic between the London Midland Region and the Metropolitan. How did the LNER and Eastern Region come into it? The LNER, successor to the Great Central, was responsible for the freight working on the Metropolitan & Great Central Joint Line and also the Metropolitan stations from Harrow to Finchley Road, and Harrow to Rayners Lane and Uxbridge. LNER steam locomotives and crews based at Neasden worked the trains between these points and Aylesbury, Quainton Road and Verney Junction on the Oxford-Bletchley line. During the war the amount of traffic dealt with from Woodford and from Verney Junction into Quainton Road was very heavy and called for special attention by the Control offices and an inspector was specially appointed on the ground. The Finchley Road exchange was very much of a relic of pre-grouping days and was a positive nuisance to the Metropolitan Railway whose prime job was to run Underground passenger trains, not freight trains which had to negotiate the third and fourth rail electric tracks at Finchley Road station. Traffic for the surviving stations on the Metropolitan with goods facilities came in from the north thereafter until it all ceased.

Tuxford was another early candidate on the list for attention. This was an historical and traditional exchange point between the Great Northern and the Lancashire, Derbyshire & East Coast Railway (LDEC) which crossed over it, with a passenger station called Dukeries Junction on both levels. There was a double junction from Tuxford North (GN) to Tuxford West (LDEC) with a reception line alongside. Coal from the collieries in the Dukeries for the south via Peterborough etc, and empty coal wagons in the reverse direction was the main traffic dealt with and for years it had been a nuisance because first it meant light engine mileage from and to Retford locomotive depot and second supervision was split, for the GN line was under the Doncaster Control and the LDEC line under the Lincoln Control. During the war congestion at this exchange was bad and in the Central Control one night, in an effort to break and beat the blockage a special coal train was set up from Mansfield to Peterborough (New England) via Lincoln and Spalding with a balancing train of empty wagons. It eased Tuxford and improved transit time and wagon user, and became an ad hoc arrangement of such success that three trains daily in each direction were established in the timetable. True, the mileage via Lincoln was a bit higher but the overall cost very much lower. The remaining traffic via Tuxford was analysed and re-routed away from Tuxford which enabled final closure to be made in 1947/48.

The same whittling away process was employed at Pindar Oaks Yard at Barnsley, a place which was on the author's Section as a Controller in 1936. Later, in 1953, at a District Operating Superintendents' Conference at Sheffield when H. C. Johnson was in the chair, a call was made for the District Officers to make some economies in marshalling yards; was it possible to close any for one shift, two shifts, or altogether? The only response was that Pindar Oaks with its small turn-over of traffic did not justify its existence and that the work could be taken over by Stairfoot which was nearby. Stephens, who was the District Operating Superintendent and later Traffic Manager at Doncaster was opposed to it, and said he was 'horrified'. Traffic analysis teams took a wagon-side record for three days and this confirmed that Pindar could be closed, but the idea was still fought by Stephens in correspondence. In discussing Stephens' attitude with H. C. Johnson he was told 'Leave it to me . . . softly, softly . . . catchee monkee' (and the file bears this endorsement!) and straightaway arrangements were made for certain traffic to be re-routed northwards via Horbury Yard, and the remainder via Stairfoot, duly advising the National Coal Board who did the wagon labelling and documentation. This took the ground from under Stephens' feet. Meanwhile, with a change in District boundaries, Barnsley, Pindar Oaks and Stairfoot were transferred from the Doncaster district to the new Sheffield district and it was left to the men at Sheffield to make the kill.

In addition to the new junctions laid in to improve the movement of war-time traffic such as the Sandy Curve (north to west from the GN to the Cambridge-Bletchley line) and its counterpart at Claydon-Calvert (Bletchley-Oxford line to the GC – effectively replacing Verney Junction-Quainton Road), and the Harringay Curve in North London connecting the GN main line with the Tottenham & Hampstead line underneath, some extra shunting facilities had been provided. Such a case was at Connington, south of Peterborough, provided to ease the shunting pressure in the down New England yards. The time had come to re-appraise them and as a result Connington Yard was closed on 5 September 1949, only a month behind the closure of Dunford.

During the war and the immediate post-war period all

marshalling and terminal yards large and small had been working flat out but traffic flows by 1950 were getting more stabilised. The closures of Tuxford, Dunford, and Conningtonton, paved the way for a general review of all marshalling facilities within the Eastern Operating Area (GC, GN, and GE) – the location of each, its sorting requirements and wagon turnover. The list in its entirety was formidable and gave cause for sober reflection and positive action. The details need not be repeated here but the total of yards in each operating district will give an indication of potential economy:

King's Cross	21
Doncaster	28
Lincoln	27
Nottingham	9
Manchester	15
Leeds	4
Cambridge	19
Norwich	2
Stratford	18
LT&S	5

The primary analysis showed that individual yards within a marshalling yard complex had been counted separately; for example Doncaster Down Decoy, Down Belmont, Up Mineral and Up Decoy were regarded as four yards because separate shunting staff and locomotives were required in each. Further, some were not marshalling yards in the strict sense but were really collecting/terminal points serving one or more local industries. The Doncaster complex of ex-GN and ex-GC yards were examined. Hexthorpe Top Yard had already been closed on 2 June 1949 and this was followed by Hexthorpe Up Yard on 9 October 1950 and Hexthorpe Down Yard on 2 July 1951, again being followed by Down Decoy for marshalling, the sidings being retained for holding or storage purposes. The Thorpe Lane Yard at Ardsley in the West Riding came under the hammer on 16 April 1951, Clayton between Bradford and Halifax/Keighley on 2 July 1951, Derby Friargate closed on 30 April 1951 and was followed by Holloway (East Goods) on one shift. All these small yard closures produced staff economies of £140,000 annually plus recovery value and no further expenditure on maintenance. On the Great Eastern

The first wagons of the trial load of coal with automatic slow speed discharge at West Burton power station in 1965.

Above View of the underside of the coal hopper wagons for Merry-go-Round trains, showing the hopper doors. *Below* The prototype 26/32 tonne coal wagon developed for Merry-go-Round trains.

a conglomerate of small yards at Temple Mills (Loughton, Blackwall, New Connection, Cambridge Yard, Colchester Yard, West Yard) gave way to a brand new Temple Mills concentration. Likewise, there was a new yard at Ripple Lane, Barking, to serve the London docks.

Browsing through personal archives in 1950 revealed a statement giving the total number of wagons detached, the normal yardstick of yard performance, at each of the main marshalling yards in 1925, some 25 years earlier which, after allowing some latitude in compilation, was rather surprising. The figures were put alongside those for 1957/58/59/60 and 1961 and, with the exception of Mansfield Concentration Sidings which served new developing collieries, all showed progressive reductions in wagons handled, as much as 50% in some cases. This was excellent material for convincing doubters when carrying out staff consultative procedures!

Plans for further retrenchment and re-organised train working were prepared for the marshalling yards at New England, Colwick, Bulwell Common, Doncaster, Harringay and Holloway (Ferme Park East Goods, Clarence Yard, Highbury Vale) Warsop and Langwith, providing for a saving of 84,000 wagon days, 82,700 shunting engine hours, 176 yard staff plus 54 other grades all then assessed at around £400,000 per annum.

The good husbandry proposal to close Clarence Yard at Holloway, putting the work onto Hornsey's Ferme Park Down Yard about two miles further north, was contested by Green the District Operating Superintendent and by his successor Christopher. The traffic analysis which had been taken showed that an average of 413 wagons were detached daily in the small dead-ended yard mainly for attachment to the eleven express goods trains which started at King's Cross and called at Clarence Yard to pick up important wagons which had come from the London Docks, and the Southern Region via Holborn and the Widened Lines, any consequential 'roughs' being filtered through to Ferme Park which was detaching 800 daily. The new train plan provided for the eleven 'stoppers' not to stop and they were reduced in number from King's Cross and a fresh train was started from Clarence Yard for Doncaster and York, giving the same forward connections as of yore, thus reducing the number of wagons which had to pass through Ferme Park. Opposition was carried on into 1957 and even 1958 when a notable meeting was held on 14 November – notable for the factual and frank memorandum which the chairman produced.

Meanwhile a work study team had been set up at Regional level under David Bowick and given the job of assessing the situation in the yards concerned at Holloway and Ferme Park – a job which was considered quite unnecessary as the available evidence was incontrovertible. But 'work study' was the new mode, the 'in' thing, and one was soon to be reading the new jargon of 'standard minutes' (aren't all minutes of a standard 60 seconds?) and 'clock-hours' (Isn't each hour the same length or is a clock-hour different from a watch-hour?). Well, the work study teams had to be found a job but to introduce them into this closure scheme was a tactical error. It proved disastrous in both principle and practice because the yard shunting staff, having put in an application for a bonus on account of increased work which could be easily contested, had been told by some-body along the line or 'they had been led to believe', or 'it had been inferred' they would get a bonus. The men's representatives said 'The employment of the Work Study team implied a bonus' and, further, 'If the shunters at King's Cross Goods have a bonus why shouldn't we?' Here we must step back a little into history. At some point in time the goods handling staff at King's Cross (the men who wheeled or carried the merchandise sundries from road vehicles across the bank into railway wagons, or vice versa) were given a goods handling bonus of so much per ton. Then, not to be outdone the shunters at King's Cross said 'We want a bonus too because we place the wagons for the handling staff to load' and a bonus was duly given, not so much on the merits of the case but on expediency. It was a bad decision. And now the shunters at Holloway and Ferme Park wanted similar treatment! One argued and proved without any doubt whatever that Ferme Park had a 50% spare capacity and in the interests of good husbandry Clarence Yard should be closed and the residue of the work be carried out at Ferme Park at no extra cost, but with a saving. The principle was maintained that if there was spare capacity at a yard, in this case 50%, it was surely within the province of management to lower it without cost or inducement, but it was conceded that if a yard was working to a high capacity and still more was required of it there might well be a case for an incentive bonus. The work study team's report confirmed the belief in a 50% spare capacity and *recommended a bonus*! Then, one asked, if a bonus were conceded what repercussionary effect would it have on economy and good husbandry schemes 'down the line' in other districts which had been carried through and others about to become operative, all without a bonus tag? Management caved in, gave the yard staff a bonus and, as predicted, it was not long before staff in other marshalling yards wanted one too!

If one may be permitted to digress, sometime between the end of the war and 1949 the author was called for an interview at 222 Marylebone Road conducted by S. E. Parkhouse and E. W. Rostern. The precise nature of the post is forgotten but it must have been up the street of personal experience. A question was put 'What do you think of a bonus scheme in connection with freight train working? I replied that I considered it immoral and iniquitous. Another interviewee was successful and perhaps small wonder in the light of thinking by top people at that time! I have never had any room for bonus or incentive schemes in the railway operating grades over and above the agreed grading system which determines the basic rate of pay; bonus schemes are a cover-up for unsatisfactory supervision and an award for doing no more than should be done, or could be done; if the payment is inadequate to attract the staff then it must be that the base rate is too low. Secondly, there is always an edge or a periphery to any incentive scheme allied to train movement, when someone on the edge of the area, or just over, may have a grievance, a legitimate grievance, and how do you deal with that? Where does it end?

The post-War economies gained impetus from the Beeching Plan and the subsequent National Freight Train Plan of 1967, the combination of which reduced all traffic flows. The wagon-load as a unit on the railway was derided – yet a lorry-load on the road was acceptable. Many goods terminals and marshalling yards and concentrations have vanished completely leaving only the odd marshalling yard like Whitemoor,

Tinsley, and Toton to continue with considerable reductions in use, having lost their main functions and surviving more as staging points. Throughout the whole system only 59 marshalling yards remained at the end of 1982.

When the Great Northern took over the Ambergate, Nottingham, Boston & Eastern Junction Railway in 1852 it included the trailing junction on to the Midland's Lincoln-Nottingham like just west of Netherfield station and enabled the GN to maintain a service between Grantham and Nottingham Midland station which incensed the Midland and led to the GN building its own station alongside the Midland in London Road (Low Level). The first junction was taken out in 1866 and from then the only connection between the two companies was through a see-saw letter 'W' in Nottingham GN goods yard, almost alongside the Low Level platforms, and it was through this that trips of freight wagons were exchanged. The Eastern Region suggested to the London Midland that this working should be replaced by through trips between Colwick and Beeston but it needed a new running junction and the London Midland thus renewed the old Ambergate Junction on 10 January 1965 – a bad move, as what it should have done (but rejected the advice) was to run the Midland into the GN and back again through straight connections at Sneinton Junction, so taking advantage of the higher level track and enabling the Midland line and three level crossings to be recovered, freeing train movements and, likewise, road vehicle and pedestrian traffic, features which had been criticised many years back as described in Chapter 2.

Improved Routing

In the early days of railways instances of alternative routes available for the forwarding of goods, mineral and livestock were quite the exception but as the railway system grew in extent and new lines constructed the availability of alternative routes increased to a state bordering on complexity. So definite instructions about routing became necessary and these were implemented by forwarding stations inserting on the invoices and waybills the routes by which consignments were intended to travel. Such routes were entered on the abstracts sent to the Railway Clearing House and served as a basis for the division of receipts. For the guidance of outside staff the route was also shown on the label attached to the wagon or other vehicle, or to the consignment. Railway Clearing House Regulation No 275 (Ed 1914) stipulated that all wagons (other than those containing mineral traffic in train loads not involving marshalling during or on completion of a journey) must have a label affixed on each side, giving among any other particulars:

(a) Name of invoicing or sending station.
(b) Destination in full and initials of the railway on which situated.
(c) Route, showing all junctions with other companies' lines and when two or more companies have lines between the same junctions, the initials of the companies whose lines were intended to be used.

The advantages of routing may be summed up in a general way as meeting the requirements of these times:

(1) The operating staff, and more especially those employed in the marshalling yards, and at exchange points, were assisted in the shunting and making up of trains, and the correct transfer of vehicles at junctions to the different companies was practically assured.
(2) The Clearing House was enabled to credit the various railways companies over whose lines traffic passed with the proper proportion of receipts due to them for the use of their lines.
(3) The accurate compilation of ton-mile and other statistics was rendered possible.
(4) The forwarding of non-common user wagons belonging to other companies in an unauthorised direction was obviated and charges for wrong sending were avoided.
(5) Traffic received the advantage of the proper train service.

Bringing the 120 or so separate railway companies under Government control during the first world war demonstrated that common sense demanded there should be an amalgamation and as we have seen this came about in 1923. Even so, it left the LNER Southern Area (GN/GC/GE) with no fewer than 111 officially recognised working exchange points with the other three companies but of these 25 were for coaching stock only, leaving 86 for freight. Commercial principles in routing persisted until natioalisation in 1948 but it was not until Line Management was set up in 1957 that it was finally wrested from the commercial department and placed under the aegis of the newly formed Freight Movement from which time operational priorities took over and which were certainly overdue.

The distance the traffic was hauled from Grimsby to Deansgate via Boston, Colwick and Brinsley Junction at Codnor Park was approximately 177 miles; from Goole via Knottingley, Askern, Doncaster, Colwick and Brinsley approximately 162 miles, compared with the direct Great Central route via Doncaster and Penistone of about 102 miles, truly a commentary on the commercial/operating outlook and railway economics of the time! This afforded an early clear example of the sort of 'round-the-world' haulage historically based on commercial ideas of securing the larger share in the division of receipts. Routing and documentation had been the academic responsibility of the Commercial Department, but after nationalisation and with the demise of the Railway Clearing House, the time had come when traffic should be worked over routes giving as far as possible the shortest haul, other things being equal; channelling or canalising traffic portended a better service for a greater volume. Under Line Traffic Management from 1957 that time had arrived and as the GN Line was astride several inter-Regional heavy traffic flows the scope was that much greater than in dealing solely with originating traffic.

The sub-section staff dealing with traffic analysis was thus sent out into marshalling yards every other week or so, largely in the industrial areas, to take wagon-side records of all arriving/departing freight trains, usually for a representative three-day period. They would then return to the office, analyse their information and present a broadsheet picture, from which, with my geographical knowledge of routes and train services, I could easily pick out the 'wrong-uns' – those wagons which were having protracted transits resulting from the old days of competition. A couple of cases may illustrate:

Bradford and other West Riding stations to Merseyside via Ardsley/Wrenthorpe, Nostell, Stairfoot, Worsborough, Penistone, Mottram, Godley/Guide Bridge was altered to via Huddersfield and Diggle

Newcastle to Derby and Burton-on-Trent via York, Doncaster and Colwick was altered to via Swinton and Knottingley (S&K) Joint, Rotherham, and Chesterfield

These cases were, of course, for individual wagon loads but specific full train loads running regularly were early candidates for revision. One was the 6.32pm semi-express beer train from Burton-on-Trent which started from the Great Northern yard at Burton and ran nightly to Newcastle via Colwick, Newark, and York; this was re-routed firstly via Bulwell and the GC route, Darnall, and S&K. In 1949 two nightly express trains of meat from Birkenhead to Newcastle via Woodhead, Worsborough, Dearne and S&K were switched to the more direct route via Stalybridge and York as also were the special trains of Irish cattle.

In another direction the old Commercial Department had a lot to answer for. For almost generations there had been, during the Cambridgeshire fruit season, a series of express trains leaving Whitemoor Yard every evening/night for Manchester, West Riding, the Midlands, York/Newcastle/Scotland, etc. The first specials despatched were passenger rated and were followed not long after by the specials which were goods rated. In most cases this created a need for duplicate intermediate and terminal points with special staff at each. Each winter it was customary for the Great Eastern and the Commercial Department to have a meeting to set up the servicing arrangements for the forthcoming fruit season and year after year from 1946 the Operating Department attended and campaigned to get a common rate and a common service to each destination and so reduce the number of lightly loaded trains the costs/receipts of which could not be viable. But the suggestion was always met with opposition from the Commercial Department which would not accept the theory that a slight decrease in the passenger rate and a slight increase in the goods rate to produce a common rate would reduce the train services by about half and maintain arrival times. No progress was made until Line Traffic Management ousted out the old departmentalism in 1957, and fused operating, commercial, motive power interests when Gerry Fiennes, the new Line Manager for the GN, and Willie Thorpe, the new Line Manager for the GE, gave it their blessing. The old thinking was broken down and in its place there was, at long last, one rate and one service – the best. But, as so often happens, it was too late; the road haulage interests had taken hold and it then became as well to let them have the lot. Beeching, the costing clerks, and the Commercial Department together lost more traffic from rail to road than they would have cared to admit.

6
COMMODITIES

Railway freight in the old days did indeed consist of many varieties, many more than the well-known '57' for they were statute bound to convey anything and everything on offer, cats or cattle, elephants or earwigs, beer or butter, hay to the City stables, half-pennies to and from the Mint, tree-trunks (otherwise round timber) to the sawmills, coal, iron ore, manure or maggots . . . the General Railway Classification book of years past gives an exhaustive list. This immense variety of traffic called for and got expertise in loading and handling even at small country stations; there was a correct way to load everything, in or on, a wagon and safely secured within the width and height gauge to ensure a transit without loss or damage, and sheeted when necessary to keep off the elements.

Some of the traffic was seasonal like the Kent fruit passing through London to the Midlands and the North, likewise a little later the Cambridgeshire fruit, the vegetables, and in the autumn the sugar beet from East Anglia and Lincolnshire. The Cambridgeshire soft fruit followed by the hard fruit season was big business and can be instanced by saying that in the 1951 season 20,787 wagons were despatched on evening express trains to Newcastle, Scotland, Merseyside, the West Riding via the ER northbound routes, and to other parts of the country via Peterborough, via Huntingdon and Kettering, via Cambridge-Bedford-Bletchley-Oxford, and via the London junctions. In the years before 1951 it was even heavier still, and it was then, in the late 1940s that a campaign was waged for a common rate for conveyance and a common service for what had hitherto gone, and was still going 'passenger rated' and 'goods rated' – a traditional Commercial Department ploy to get a higher rate for at least some of the traffic (the old maxim of 'what the traffic will bear') but without due regard to the extra cost in maintaining duplicate services, each underloaded. The matter was raised annually at the fruit train planning meetings but without avail until the early Spring of 1958 when Line Traffic Management was 'in' but, as mentioned in the previous chapter it was too late and the traffic was lost to road haulage.

Vegetables in the raw! At one time there was a Potato Market at King's Cross Goods with rail and road access and with the merchants' premises in between. Every night three train loads of potatoes arrived from South Lincolnshire (Algarkirk) and the Cambridgeshire fens (Long Sutton), the wagons being positioned in the respective bays allocated and rented by individual merchants.

Rhubarb? There were two express freight trains every night from the West Riding of Yorkshire known as '904' and '906' the latter being the extra to convey the rhubarb for London and southern counties. Both ran fast up the GN main line to East Goods and King's Cross as also did the Green Pea Special which started back at Selby at 7.10pm attaching at Doncaster green pea traffic from the Goole and Doncaster environs, leaving Decoy at 8.36pm and arriving at King's Cross Goods depot at 1.40am ready for the morning wholesale markets at Covent Garden and Spitalfields.

Between Christmas and Easter the seed potato traffic from Scotland was large enough to justify a big programme of nominated trains to run to selected points in Lincolnshire and East Anglia for distribution. In some years, following bad weather in Scotland, the early despatches were delayed and later intensified the movement problem by compressing more trains into a shorter period up to Good Friday which was the traditional planting time by countrymen in the south.

Sugar beet? No fewer than 13 of the 18 sugar beet factories were situated on the Eastern Region – Felstead, Ipswich, Ely, Bury St Edmunds, Wissington, Cantley, King's Lynn, Spalding, Peterborough, Newark (Kelham), Brigg, Bardney and Colwick – so there was a tremendous movement in wagons from country stations under a permit system issued by the British Sugar Corporation, and also raw sugar to the refineries. An LNER Eastern Region officer was the official liaison between the whole rail system and the Sugar Corporation. The latter knew the likely tonnages from each growing area and movement was planned as far as possible to mutual satisfaction but there were occasions when there was a build-up for a particular factory necessitating diversion to another one. The beet harvesting season, known in BSC circles as the 'campaign' usually extended from mid-September to mid-January but actual opening and closing dates at the factories were subject to variation governed by crop conditions and quantity.

Once started, production proceeded continuously until the end of the season and each factory had to have at all times, day and night, at weekends, and during the Christmas holidays, a supply of beet sufficient to sustain maximum production. While the major portion of the crop is grown in East Anglia and Lincolnshire each factory is situated in a growing area from which it normally draws its main or initial supplies. But it follows that a very large proportion is grown within a comparatively short distance of the factory. There is, however, a considerable acreage of beet grown in the south of England – from Kent to Cornwall – referred to by the BSC as its Southern Area, and as there is no sugar beet factory in this area the whole of the tonnage has had to be sent to Kidderminster, Allscott near Shrewsbury, Colwick, and occasionally to Peterborough. These long hauls were almost exclusively by rail which accounted for a higher proportion of rail beet being received at these factories. Wagons used for the conveyance of beet were coal wagons which, after being emptied at country stations, would normally be returning empty to collieries. Beeching's closure of branch lines and country stations cut out the carriage of domestic and industrial coal and, likewise, the sugar beet. It was estimated that about 60% of the beet crop was drawn from within 20 miles of the factories and 90% within 30 miles so it was a gift to the road haulier who could return with a load of beet pulp for the farmer to use as stock feed, or with a load of good clean fenland soil which had been washed from the beet. Of the 18 BSC factories, 13 produced white sugar and five (Bury, Ely, Peterborough, Spalding and Wissington) raw sugar, a considerable tonnage of the latter being despatched by rail in bulk to Tate & Lyle's refinery at Silvertown. For these rail carryings, 16-ton steel mineral wagons, specially caulked and fitted with sheet supporter bars, were placed in a shuttle working between factory and refinery.

In the 1953/54 campaign 128,985 wagons were loaded at BR stations of which 82,405 were from the then Eastern Region – GN, GC and GE – constituents. During the 1959/60 season the Colwick factory accepted the largest amount – no fewer than 15,102 wagons but subsequent years showed progressive declines.

Fish? This is a subject which many ex-railwaymen reflect upon with considerable distaste – distaste at the way in which it was handed over to road hauliers, as it were, on a plate. Fish really was big business. Taking 1949 as a representative year for a nation-wide review 1,001,625 tons were landed at British ports, about 75% of those on the East Coast in what had been LNER territory. The bulk of it was carried by rail in some 7,000 wagons each week and in addition small consignments were sent from the ports in parcels vans attached to ordinary passenger trains; many such boxes or bags of fish despatched by wholesalers at Hull and Grimsby on the afternoon passenger trains were collected by the consignees in the late afternoon or early evening and used in the fish and chip restaurants and shops that night. The Humber ports, in particular, were the despatchers of huge quantities on express fish trains which were treated with respect and great pride. From Hull there were seven trains, 12.30pm to Banbury, 3.30pm Plymouth, 3.32pm Holloway (East Goods) and King's Cross, 6.5pm and 6.45pm Banbury, 6.55pm King's Cross, and 7.42pm Manchester (Guide Bridge). From the railway-owned docks at Grimsby and New Clee there were the 1.4pm Banbury, 4.30pm Cardiff and Whitland, 4.48pm Nottingham, 5.13pm Manchester (Ashton Moss), 5.30pm and 5.55pm Holloway (East Goods) and King's Cross, 6.13pm Cambridge, 6.25pm Leicester, 7.0pm King's Cross and the 8.25pm to Leeds and West Riding. All these trains took fish for the destinations shown and for a wide range of distribution. Each night two trains ran from East Goods to the Southern via the Widened Lines. From Scotland came the 12.30pm daily from Aberdeen due into King's Cross at 3.5am and this was sometimes followed by a second train leaving Aberdeen at 1.40pm.

Loadings per train naturally varied with the season and the weather at sea but they were generally heavy. An occasion is recalled when consideration was being given to the uprating of the East Lincoln line between Peterborough and Grimsby to improve the locomotive route availability with strengthening of the permanent way to permit the use of more powerful locomotives. In dealing with the economics of the proposal the gross, tare and net weights were recorded of the 7.0pm Grimsby to King's Cross and it was found that the locomotive was hauling at express speeds of 45-50-55mph a trailing weight heavier than a train load of coal running at less than half that speed; the fish vans were loaded to the hatches, as it were, with wet iced fish. It is worthy of record that the highest number of vans of fish received into King's Cross Goods depot in one night was no less than 274. The fish trains had long hauls; for example 539 miles from Aberdeen to King's Cross with a two-day van turn-round time.

In 1964 the transit of dead fish was very much a live matter, with the Eastern Region fighting for its retention against the Board headquarters which had a bee in the bonnet that fish did not pay its way. That was nonsense, unless one agrees that any traffic treated in isolation did not pay. What if the post office and the supermarkets adopted the same principle and jettisoned the individual loss-makers? Each treated in isolation would result in a business closure. The Work Study and the Costing Service arriving at arbitrary decisions born out of inexperience have a great deal to answer for. In spite of Regional opposition the Board won the day over the transit of fish and handed it over to road hauliers – ironically at a time when the Board itself was preaching the gospel of traffic movement in train loads (in which much of the fish travelled) rather than in wagon units! And what of the fish docks built by the railway, owned and financed, and laid out track-wise for rail servicing? Instead of a number of trains of empty vans all marshalled in orderly fashion alongside the quay and fish market for loading by the wholesale merchants, and prompt despatch in a train load in a few minutes, what is there now? Hundreds of road lorries and lorry drivers in a melee, to say nothing of the congestion in the streets approaching the docks and in other urban areas and certainly with journeys no quicker than rail. But is there still a market for fish, for wet fish shops have declined?

Meat? As with fish so with meat. The 10.12am express meat train from Aberdeen to King's Cross, heavily loaded, arrived regularly, punctually like a passenger train, into King's Cross Goods terminal at 1.12am a time eminently suitable for the early Smithfield market, or elsewhere. But that too has gone.

Bricks? The output of bricks from the Fletton brickyards at Peterborough was phenomenal, with large quantities of empty wagons from the south being fed into the sidings adjacent to the kilns. The annual tonnages were around 1¼ million. During the second world war nominated loadings of full trains were arranged for via London services, to Scotland, Doncaster, Hull, York, Birmingham and East Anglia; for the latter even the Huntingdon North/East – St Ives route was brought into use.

It was the practice for train loads for the Birmingham area to reverse at Peterborough North and again at Peterborough East, the locomotive running round the train in each case until, at the suggestion of the author, the old Longville Junction on the ex-LNW line west of Peterborough East was opened up again. The junction connection was originally opened on 2 July 1883 and was used by the Great Northern passenger trains from Leicester (Belgrave Road) to enable them to get into the Great Northern station at Peterborough, afterwards designated Peterborough North, joining the East Coast main line at Fletton Junction. The passenger service was abandoned on 1 April 1916 as a first world war economy measure and was never restored. The points and signals were taken out and the track cut, the remainder of the branch to Fletton Junction being used for servicing private sidings which included the sugar beet factory. The restoration of the Longville Junction was readily agreed and was functioning within months,worked by a ground frame, the installation cost being £3,000. From then the London Brick Company provided a daily train from Hatton, and the Fletton pilot propelled it down the Botolph branch to Longville Junction at which point the engine for forward working was hooked on.

Alas, by 1959 the annual rail tonnage of bricks from Fletton had dropped from 1¼ million to 354,954, and in more recent years down to 100,000 mainly to railheads at Billingham, Gateshead, Sheffield, Leeds (Stanningley), Hull and King's

Cross, a much lower figure partly due to building recession and stockpiling. Facing bricks which require minimum handling, form 50% of the output and a Freightliner service to King's Cross and the North West has been built up.

Steel? Reference has been made elsewhere to the steel industry which provided so much traffic to the LNER as a whole and what it meant to places like Scunthorpe, Staveley, Ilkeston (Stanton), Sheffield and the Don Valley generally but there was much more of an involvement with associated commodities. Iron ore was mined on the Denton, Eaton, Highdyke and Stathern branches, that from Highdyke reaching five and six trains per day. There was also a mine at Holton-le-Clay on the Lincolnshire Wolds near Market Rasen, and in earlier days there was daily train from Caythorpe between Grantham and Lincoln to Stanton Works via Colwick. Iron ore also came from other Regions – from Ashwell, Edmondthorpe, Market Overton and the Kettering area, and through the port of Immingham there were large cargoes from Africa, Brazil, Spain and Scandinavia. Other industries closely linked as they provided the essential ingredients for iron and steel production were coke, slack, lime, limestone and scrap iron. The chief firms at Scunthorpe were John Lysaght, Firth Brown, Richard Thomas, and the Appleby-Frodingham Iron & Steel Company the latter being a vast complex covering many acres and as mentioned in Chapter 2 enjoying no fewer than five entrances from the railway into the works, traffic being appropriately labelled by senders to Entrance A, B, C, D, and E. according to consignees' requirements. Within the orbit of railway operation blast and steel furnaces were situated as under:

	Blast	Steel
Scunthorpe	10	26
Renishaw	2	
Sheepbridge	2	
Staveley	4	
Parkgate	2	10
Chesterfield		2
Rotherham		3
Sheffield		145
	20	186

All along the Don Valley were the works which transformed the pig iron and steel ingots into railway wheels, axles, tyres, spades, and all sorts of tools, cutlery and precision instruments. Names like Steel Peech & Tozer, Parkgate Iron & Steel Company, Firth Brown, English Steel, Brown Bayleys, Davy United, and Cravens, all firms of substance and repute, all household names in that part of the world and far beyond; they were the railway's clients. And so, of course, was the Stanton Company at Ilkeston fed with iron ore via Colwick.

Above The lineside equipment for opening wagon bottom doors as the wagons reach the unloading bunkers at West Burton power station. *Below* Coal being discharged through the grills into the hopper below at West Burton from wagons moving at dead slow speed.

The first trial train pulling slowly away empty after discharging its load at West Burton power station.

7
The Birth of TOPS

For some years before setting up the Line Traffic Management organisation in 1957 the Operating Department had a strong liaison with the Goods Commercial Department in endeavouring to meet traders' wishes regarding the transit of their consignments, large and small. This was intensified so that when the new organisation did come in it would be well primed and imbued with a desire to do even better for customers, consistent with the economics of a particular service. And one way in which the Eastern Region and, indeed, its predecessor the LNER, tried to effect fool-proof transits was the 'wiring-on', that is telegraphing forward of important and perishable consignments by the wagon load and applying more particularly to the night express freight trains. The individual railways had a very extensive telegraph system linking every station and depot, and Great Eastern lines tended to have telegraph instruments in offices only but, as well as telegraph offices, the Great Northern also had them in signalboxes. Every signalman on the Great Northern had to learn, receive and send in Morse. Some offices at important junctions had as many as a dozen clerks on duty at a time receiving, forwarding and re-transmitting telegrams on railway business (or in some cases postal telegrams acting as agents for the Post Office). The railway telegrams were usually and partly in code and here are just a few examples of the many in the code-book, and their meaning:

GUM	Your wire of today
GOLO	Your wire of yesterday
ARROW	Reply by wire on the following subject
FONET	Referring to your telephonic communication
WALNUT	Make all necessary arrangements so far as you are concerned

OHIO	Send on all speed
COG	Accommodation has been reserved
FLEA	Wire how you stand or what you are doing
COW	Be prepared to detached undermentioned vehicle leaving here in the rear of the following train

There were hundreds more covering almost any circumstance or describing any type of vehicle.

Over the years, however, the growth of the telephone which anyone could use, as distinct from having to learn the Morse code, and with it the growth of train and traffic control, changed the pattern of communication and much that had hitherto gone by telegram was transmitted by teleprinter or telephone and/or within limits, via control offices. To a point that was right, for a controller had some executive authority which a telegraph clerk had not, but the earnestness to please and the reliability overloaded the Control system of passing train loading information from Control to Control and finally to the terminating marshalling yard or depot. It was found that some Control office staff were becoming telephone clerks to an unjustifiable and unacceptable degree and were being swamped. For instance, the 3.35pm through express freight from King's Cross to Edinburgh and Glasgow, consisting of 50 wagons, was advised forward in this sequence:

King's Cross Goods depot
King's Cross Control
Doncaster Control
York Control
Newcastle Control
Edinburgh Control

as '50 Scotch including . . .' and there followed a long list of miscellaneous types of traffic such as ice cream, motor cars, household removals, etc, for various destinations in Scotland but the only point of interest in the detail composition of the train was in Scotland for the train was not touched until arrival at Edinburgh (Niddrie). Arrangements were made in this case for King's Cross Goods depot to telegraph or telex the detail to Scotland direct and for the information Control to Control to be limited to '50 Scotch'.

But this was only the tip of the iceberg and fortuitously in the late autumn of 1954 the head of the Eastern Region telecommunications section rang to say the British Road Service headquarters had some teleprinters to spare and could some use be made of them? The campaign with all the District Officers against the gross over-loading of the inter-Control train advices was already under way and this offer could not have come at a better time, for the idea had already been conceived of having a teleprinter in a cabin at the departure end of the major marshalling yards with a specially appointed clerk to record the wagon-side details of a formed train and then proceed to his cabin and teleprint the information to the terminal yard or depot, or a summary to intermediate Controls; all the latter had to do was to tear off at the perforation mark.

On 2 and 3 December 1954 a meeting of Chief Controllers and Chief Train Clerks from King's Cross, Lincoln, Doncaster, Nottingham, Leeds and Manchester, was held at Sheffield when the opportunity was taken of bringing this possible development to their notice ready for a special meeting it was proposed to call at Liverpool Street on 22 March 1955. The latter was a memorable meeting in more respects than one for by this time H. C. Johnson had become Operating Superintendent of the whole of what was known as the Southern Area of the Eastern Region comprising the Great Central, Great Northern and the Great Eastern, and the author had to assume responsibility likewise for freight planning and operation. It was the first time ever that a meeting had been held on the freight side with representatives from all ten ER districts. At this meeting there was overwhelming support for the idea and it was decided to operate a pilot scheme between Colwick and Whitemoor marshalling yards. Moreover, the American term 'train-consist' was introduced and it became a part of the terminology instead of 'train loading'. Various types of pocket tape machines were tried out by four Traffic Analysis staff for use instead of book and pencil records but were not sufficiently reliable. Unfortunately for the Eastern Region the word got around of its intentions. The potential was recognised and further progress was officially halted, for the BTC wanted to study the Inglis plan of wagon-side photographic recording by lineside equipment photo-cells as trains passed by. Two approaches were made to firms specialising in the electronics, but the plan was ruled out by the exhorbitant cost, apart from certain inherent snags.

In September 1957 the departmental organisation was disbanded and the Operating Superintendent, the Goods Commercial Manager, and the Passenger Commercial Manager, each responsible for the whole of the GC, GN and GE forming the Southern Area together with their respective staffs werere-organised under Line Traffic Management with the three departments fused under Gerry Fiennes for the Great Northern Line (GN and GC) at Great Northern House in Euston Road, and under Willie Thorpe for the Great Eastern Line (GE and LT&S) at Liverpool Street. The newly-formed Freight Movement sections of both Lines, tired of waiting and having wasted three years, picked up the pieces again. The GE Line operated an internal pilot experimental scheme between Whitemoor and Temple Mills yards. On the Great Northern Line there were further meetings with the Districts, newly designated as Divisional Traffic Managers. On 5 May the author, with others, had a discussion with Ferranti Ltd with a view to using the spare capacity in the computer which had been installed in Great Northern House, primarily for the time-tabling of trains. There was a clear idea of what was wanted – to process every wagon movement in and out of marshalling yards, stations and depots, so affording wagon control date along the line and progress in other ways from there. Unfortunately the computer was withdrawn. Nevertheless, the Great Northern Line succeeded in setting up a system which was dubbed Advance Traffic Information (ATI) at all major yards, with train preparers and guards who had spare time at the beginning or end of a shift, to take the records, using tape recorders, teleprinters and telex. It was successful and the system was subsequently extended up the East Coast network embracing the former LNER system in the North Eastern and Scottish regions. The BRB set up an organisation at the Marylebone Road headquarters to develop the idea nationwide and computerise it (denied to the Eastern) with the two large IBM370 computers being christened 'TOPS' – Total Operations Processing System – and giving a countrywide freight information service over 400 terminals. Authorised in 1971 it was initiated in the West Country in 1973 and extended 'all-line' in 1975, 20 years after conception by the Eastern Region. Valuable as the sophisticated TOPS is, the tragedy is that its implementation was so markedly delayed that the cost of it increased by inflation to £16½ million and the benefit derived from it is in inverse ratio because the traffic volume has gone down as already discussed from 283m tons in 1954 to 194m tons in 1973 and, further, to 153m tons in 1980. The system did however prove a boon in public relations in as much as many major companies are now linked directly to the TOPS computer and can obtain current information about their own flows of traffic.

All the ingredients of the TOPS pudding were prepared by the Great Northern Line Management team and it is rather disheartening that the inspired forward thinking, action, and team-work, was recognised *only* by *The Railway Gazette* of 19 August 1966: 'The improved transit information arrangements which are being incorporated into the National Freight Train Plan were inaugurated on the Eastern Region and comprises a system of advance traffic information between marshalling yards' . . . and further . . . 'the seeking of an efficient system of gathering information on the make-up and content of express freight trains provoked a great deal of thought. Various suggestions were examined on the Eastern Region and the current system is a direct result of these.'

8

RE-ORGANISATION: LINE MANAGEMENT

The Railways Act of 1921 reduced and fused 120 separate railway companies into four groups, Great Western, London Midland & Scottish, London & North Eastern, and the Southern, each with its General Manager or equivalent; one may say 'equivalent' as the LMS always had to be different. Each had a natural geographical territory with an operating 'line of route' but where there was inevitable overlapping provision was made for discussion on likely economies from the pooling of resources or the withdrawal of a company from a particular locality. Rightly so, but unfortunately a good deal of traffic was lost from rail to road; the goodwill and business built up by local officials and traders were overboard, the redundant staff were absorbed or transferred possibly to an alien company and with somewhat bitter feelings. This well-merited organisation was the forerunner of a series of disorganisations and reorganisations in one form or another and in different parts of the country.

The Transport Act of 1947 established the British Transport Commission as a public authority with Lord Hurcomb as the first Chairman until 1953 when he was succeeded by General Sir Brian Robertson, – the administration and development of the railways and other transport services being delegated to five public bodies known respectively as the Railway Executive, London Transport Executive, Road Haulage Executive, Docks & Inland Waterways Executive and Hotels Executive, and on 1 January 1948 British Railways came into being with an adjustment of regional boundaries clearly defined to suit the Commercial Department but which automatically provided 'penetrating lines' for the Operating Department. The Railway Executive was the main concern, the Chairman of which was Sir Eustace Missenden; he had six

The split of routes for Line Traffic Management on the Eastern Region in the late 1950s, which provided independent management at Line Level for the GN, GE and London Tilbury & Southend lines.

full-time functional members. The railway system was divided into six largely geographical regions, each with its own headquarters staff of departmental officers responsible to the appropriate functional member of the Executive. But to ensure co-ordination of all departments with individual regions and that all staff worked as a team, a Chief Regional Officer was appointed in each Region, exercising supervision over it as a whole on behalf of the Railway Executive. For a departmental officer to communicate and report directly to the functional member, by-passing the Chief Regional Officer, brought difficulties which should have been foreseen.

In the Transport Act of 1953 the Railway Executive ceased to exist as from 30 September 1953 when its functions, together with those of the other Executives, were assumed by the British Transport Commission. Under the terms of this Act the Commission had to submit to the Minister of Transport a scheme for the re-organisation of British Railways incorporating decentralisation of management while reserving to the Commission general financial control and general control of charges. This scheme provided for the introduction of Area Boards and came into effect on 1 January 1955 when Sir Reginald Wilson went to the Eastern Region as Chairman of the Eastern Area Board. He was one of six Area Board chairmen, not one of whom was a railwayman. It was Sir Reginald who did away with the time-honoured, departmental organisation and set up Line Management, first with the London, Tilbury & Southend line as an experiment, and then with the Great Eastern and the Great Northern. But before his departure in 1960 he had set up a committee of three which, as prophesied, unscrambled it.

The Chief Operating Superintendent of the Eastern Operating Area for a short time was H. C. Johnson, with G. F. Fiennes as his Assistant and with J. W. Dedman and John Bonham-Carter as 'Assistant to'. In July 1957 'HCJ' became Assistant General Manager and Gerry Fiennes succeeded him as Operating Superintendent of the Eastern Operating Area as distinct from Eastern Region, by reason of 'penetrating lines', bringing together the GC, GN, and the GE beneath one operating umbrella. But even this variation did not last very long.

Very early in 1957 a flurry of meetings was the initiation to another move, reorganisation of regional boundaries and the elimination of penetrating lines and, with it the setting up of 'Line Management' on a wider scale. The LT&S Line from Fenchurch Street to Tilbury, Southend and Shoeburyness had already been going as a pilot scheme under J. W. Dedman as Line Manager. The GE Line covering East Anglia was to comprise the Liverpool Street, Norwich and Cambridge districts with Line Headquarters at Liverpool Street and with operating, commercial and motive power departments fused at all levels. The GN Line headquarters was to be at King's Cross in a new block of offices christened Great Northern House in Euston Road, to supervise and co-ordinate departments in the King's Cross, Lincoln, Doncaster, and the new Sheffield district which also contained the old LMR Rotherham district territory. Lines operated by one Region which had become penetrating lines into another geographical Region had to be regularised and this meant that as from 1 February 1958 the Eastern Region lost the Great Central main line from Marylebone via tht GW&GC Joint and via the Met & GC Joint, through Woodford Halse, Leicester, Nottingham, and Annesley as far as Pilsley, also from the Dunford portal of Woodhead tunnel to Manchester and beyond as well as the GN lines at Colwick, Nottingham and to the west, that is to the Leen and Erewash Valleys. All were taken over operationally by either the Midland or Western Lines of the LMR at Derby and Crewe respectively. Likewise, the GN-operated lines in the West Riding north of Doncaster and the GC lines north of Barnsley to Wakefield, Leeds, Bradford, Keighley and Halifax passed over lock stock and barrel into the control of the North Eastern Region at York.

The Eastern Region's gain of the Midland main line from Chesterfield (Horn's Bridge) through Sheffield and Rotherham (Masborough) to Darfield and Cudworth, that is, the old Rotherham district, was no recompense. As the Midland main line from St Pancras to the East Midlands lay geographically between the GN main line from King's Cross to the North and the GC main line from Marylebone, and having regard to junction connections one with the other a good case could no doubt have been made out for the Midland main line to have been transferred to the Eastern as between St Pancras and Chesterfield.

Great Northern House staff were appointed as from 16 September 1957 and took possession of Great Northern House just over a month later in one weekend. With the strict operation of Regional boundaries following the elimination of penetrating lines went a re-shuffle of District Offices and in particular the boundaries of District Control offices. The Eastern Region lost the Manchester, Nottingham, and Leeds Districts and Sheffield in effect took the place of Rotherham which it gained. It all entailed a tremendous amount of planning, accommodation, staffing, telephones, the latter being a major task so far as the Central Control and the District Controls were concerned.

The erstwhile departmentalism fused beautifully under the new more-embracing umbrella of 'Traffic Management'. It was a remarkable transformation from past days of petty jealousies, bitterness, acrimony and, maybe distrust, a composite feeling conditioned and governed by departmental heads of years before. If two or three of the latter were at loggerheads or nearly daggers-drawn – or with the mere existence of a 'feeling' – then it permeated right down; the Loco-man searing at 'the bloody Traffic', or the Traffic man swearing about 'the bloody Loco'! The author has heard a self-opinioned swollen-headed clever-dick of a District Officer say when about to participate in the morning telephone traffic conference – a sort of round-table conference with all district officers at the same time 'chaired' by the Operating Superintendent or his Assistant – 'Let's see what these silly b's want now?' That outlook ran right through his office from top to bottom because of the example that was set.

Often it was brought home to me in early days how much a departmental chief or office chief influenced his staff, and that he and they seldom realised it. District Control offices were, for instance, like school class-rooms. The deputy chief controller in charge of each shift was like a teacher; if he was quietly efficient you could hear a pin drop, for all the controllers were quiet, but if he happened to be an excitable type of man (in which case he was wrongly appointed to a Control Office) or a 'loud-mouth' then the office, like the class-room would be a babble of noise. Loud talkers in Control Offices were often 'show-offs'; they wanted others to hear the action they were taking.

At Great Northern House the chief, Gerry Fiennes, his functional assistants and their respective staffs, formed the happiest and most efficient team of management, and many have testified to this. Proof was not only manifested in personal attitudes but fully evident in the conglomerate of statistics. Gerry Fiennes had ability and inspiration of the highest order. He had 'gone through it' and was a manager who had never lost the common touch or forgot the christian names of the many who had passed through their hands over the years.

As a year or two passed changes in management personnel were inevitable and in one way regrettable as it broke the initial links which had been forged but, fortunately, successors caught on to the prevailing spirit and it was generally the younger end who went on to other appointments in or outside London. Nevertheless the family feeling engendered and firmly established by Fiennes carried on when he left us for the Board HQ as Chief Operating Officer. Fiennes' departure in 1961 was a great loss to the Line Traffic organisation for he was a man of stature with a considerable welding cohesive ability which showed itself not only in the work-life of his staff of all grades both inside and outdoor but also in their off-duty pursuits; he did not demand loyalty for it was freely given. True, 'GF' set off a few hares, and schemes were worked up to some degree, notably reorganised marshalling yards at Colwick, New England and Doncaster, all in good faith but which because of their elaborateness were almost dead ducks from the start. But he had an earnest desire to do another Stratford/Temple Mills reconstruction but too much time had passed in between and the climate had changed: by the early 1960s the 'in' policy was that of Marples and Beeching and the Government's attitude veered in favour of road. As things turned out traffic-wise it is a good thing that those projects in the air never got any further.

There were other changes but the pattern set by Fiennes was accepted and adopted by newcomers in those years which were agog with happenings, some good, some bad, some inconvenient. At least four were in the latter category as they had the Freight Assistant leaving home at bedtime or in the small hours to spend the rest of the night and next day in the Central Control as Duty Officer – a circumstance brought about by serious derailment or collision.

On 19 November 1958 there was a collision at Hitchin at 4.29am and such a procedure was followed. One freight train had run into another at Hitchin South in fog and, as a result, the engine and a number of wagons were derailed. Unfortunately,the 10.35pm express freight from Leeds to King's Cross travelling on the up main line ran into the wreckage, derailing the engine and wagons which, in turn, hit a light engine standing on the up slow line and which was flung over on its side. All four lines were blocked, the first being cleared at 11am and the other three at 5.45, 7.5 and 8.5 the following morning. As the collision was between Cambridge Junction north of Hitchin and Langley Junction south of Hitchin, it effectively shut all expresses, outer suburban, and freight trains out of King's Cross and adjacent marshalling yards and called for wholesale diversions to be arranged via Peterborough East and Cambridge and also via Sandy and Cambridge into Liverpool Street and Stratford. There also had to be diversions of passengers away from King's Cross to travel northbound via St Pancras or Liverpool Street. From the operators' view point there were the attendant difficulties of engine route availability, drivers' road knowledge, pilotmen, and the unbalancing and/or reversing of coaching stock. Important long distance express freight trains were diverted via Harringay Junction, the Tottenham & Hampstead line and the Midland route, going back to the Eastern Region at Chesterfield (Horn's Bridge); other freights were held back until the line was cleared. It was a full scale operation which entailed long hours in the Control Office.

On 15 December 1961 there was a collision at Connington a few miles south of Peterborough, on 7 June 1962 a bad derailment at Lincoln and on 7 September 1962 a collision at Offord, between Hitchin and Huntingdon. Such occasions as these are naturally highly regrettable but one would be far from honest not to confess revelling in the challenge they presented, for the many war-time emergencies dealt with in the Central Control as Deputy Chief Controller provided a good training for taking things in one's stride.

When accidents like these happen there is no place for anyone who is easily excited, or timid; there needs to be a quick appraisal and an ability to make speedy decisions. Doctors and ambulances if necessary are called by the District or Divisional staff or today the area managers staff if such action is necessary, but at headquarters arrangements have to be made for adequate cranes and breakdown equipment to attend by withdrawing some from other districts, clearing any trapped trains, if necessary pulled back from the dead line, and diverting them over other routes, possibly through other regions, subject to locomotive route availability and the knowledge trainmen had of such alternatives, pilotmen being provided as required. These occasions, although arising from adversity added spice and zest to the job and at the end some satisfaction with the result.

There was one occasion in the 1934/36 period in Central Control when as a young Assistant in charge the East Coast main line was totally blocked in the middle of the night, and likely to be so for some unforeseeable time, at the Sandy bottleneck where four lines were reduced to two through the station platforms. Several sleeping-car and other express trains were trapped north of Sandy and had to be hauled back to Peterborough North and diverted via Peterborough East, March, Cambridge and Hitchin. It seemed that a second breakdown crane was desirable so I ordered the one from Peterborough to come out and work at the north end while the King's Cross crane covered the south end. There was some local district criticism of my instruction but Barrington-Ward defended my action and decreed 'never hesitate in the case of a main line derailment to get a second crane out if you feel it necessary even if in the end it is not used'. This was a maxim drummed into Controllers in after years.

The route mileage managed by the newly-formed Great Northern Line Organisation in 1957 was around 1,500. There were about 10,000 originating freight trains weekly with a freight train mileage of about 360,000, discounting the through inter-Regional trains of which there were very many. Track mileage was about 4,000 with ome 350 freight stations/

depots working. That was the background to the day's work for the senior managers.

One could say work started in the morning by doing this or that but on reflection, it was never regarded as work in the sense that work is a 'grind' or a chore that has to be got through somehow. Probably this was due to the fact that it was absorbingly interesting insofar as one could use initiative, organise, plan, monitor, and carry out a post-mortem; then one could reflect to see if it was a success or a flop. However, work or duty started by going through the Central Control log for the previous 24 hours, noting the incidents and the decisions taken by the Deputy Chief controllers covering the three shifts, interpreting a broadsheet giving a summary of wagons on hand and in the pipe-line in each District for all main routes, seeing what the motive power availability was like and what adjustments might be necessary, and scrutinising the running sheets of the express passenger and freight trains. More often than not the Freight Assistant took the chair at the morning round table conference on the special telephone line with all the Divisional Operating Managers or their Assistants, listening to their reports, suggesting this or that, co-ordinating traffic and engine power dispositions, criticising when it was thought necessary on matters relating to punctuality, or delays in dealing with mishaps or accidents.

Preparation for and taking the conference took about 1½ hours and it was a very invigorating start to the day; it was inevitably so.

Compared with some officers (here today, gone tomorrow, never staying long enough to catch the rabbits they had let loose!) there was a distinct advantage in conducting the morning traffic conference with confidence because if the chairman had not only worked in each Division but had travelled extensively over all lines and particularly in the complex Notts/Derby/Yorkshire coalfields he knew what the people at the other end of the telephone should be talking about. Thus the location of junctions, general layouts, yard and station capacities were almost always readily pictured in the mind and, generally speaking, signalboxes and the lines they controlled with their different nomenclatures were well known. For example, at Doncaster the up fast line was not the up main from Bridge Junction to Red Bank and Decoy; the up fast ran from Bridge Junction to Carr Box and Decoy. Another example, the difference between New Oaks Junctionn and Old Oaks Junction at Stairfoot, they are two signalboxes on two separate triangles. Sometimes at the morning conference a District Officer could be reticent over an incident or in explaining a delay which had been questioned, or he might try to pull a fast one hoping it might come off! One District Officer when asked how matters were in his district, had a habit of 'quick-fire spouting' a string of trivialities hoping to take up time to the point of weariness and create a diversion from other aspects which merited greater attention. (Sometime after retirement the author was paid a compliment by two officers still serving; they said 'We were talking about you after the conference this morning when so-and-so pulled the wool over the eyes of the inexperienced HQ officer taking the conference and we both said "My word, Reg Munns wouldn't have let him get away with that one"').

After the morning conference there was the general supervision of the Freight Wagon Control, the Central Train and Traffic Control and the Freight Movement Section. The latter had a number of sub-sections dealing with development, transits, traffic analysis, routing and labelling, and a most specialised section which looked after the movement of out-of-gauge and exceptional loads. Ralph Wass was the king-pin of this aspect on the GN line and it is doubtful if any railway, or railway region, at any time since 1825, had his equal in this field of operation. Basically, he was an operator but for years had had the full confidence of the commercial rates and charges staff; he had a way of finding out what alternative forms of transport would charge to enable him, after consultation with the Civil Engineer's and the Signal Engineer's departments when necessary, to quote a price for the job; this included providing the right type of vehicle, loading to left or right of centre, possibly adjusting one way or the other to avoid a structure en route, and craneage, be it a transformer, a ship's propeller, or bridge girder work of which the ER conveyed a lot. On occasions he was loaned to the Continental Department to oversee the transit of some large unwieldly consignment on ship and forward on the French/Belgian/

German railways. Incidentally, the heaviest load carried by British Rail was arranged by the Eastern Region in September 1968; it was a 122ft long boiler drum from Immingham to Killingholme.

For very many years, even from early railway days, it was the practice for colliery companies (and later the National Coal Board) and traders generally to communicate through the railway commercial department in the District or at Headquarters level, and the commercial department then had, of necessity, to confer with the operating and/or engineering departments. Traffic Management cut out all inter-departmental lines of communication; after all, the commercial department could only quote the rate and do the documents. The commercial department was quite unable to negotiate on their own details of private sidings layouts, methods of working, transits and servicing generally. Many big concerns such as Shell-Mex, the Central Electricity Authority (later CEGB), National Coal Board, Stephenson-Clarke, Charringtons, the British Sugar Corporation and the British Iron & Steel Federation, were not slow in realising this. They refused to conduct their discussions with anyone but those directly responsible for movement. 'Traffic Management' gave this facility and an excellent relationship ensued.

On the LNER, if not generally, it had always been the rule that the timing of trains and consequential locomotive and crew diagramming was to meet the commercial and operating requirements of the freight and passenger sections, but not overlooking the economic user of manpower and the hardware of locomotives and track. Indeed, this was stressed time and time again in words from the top '. . . the Central Timing & Diagramming Section is the servant of the Trains Section'. Accordingly, in the Line Management organisation the Timing/Diagramming/Timetable section was under a Head who was immediately and jointly responsible to the Passenger Assistant and the Freight Assistant.

The Line Management organisation instituted in 1957 with the Great Northern Line proved so effective and efficient that, in effect, it contributed to its own demise, for in the 1960s it was decided virtually to amalgamate the lot in a new organisation under the General Manager, cutting out one link in the

Left Train of empty hopper wagons leaving West Burton and passing the visual examination point where checks are made to see that the hopper doors have closed.

Below Trial run with Merry-go-Round hopper wagons heading for West Burton with the LNER dynamometer car next to the locomotive. (Eastern Region, BR)

Above The now closed electrified route between Manchester and Sheffield/Wath. Two Class 76 electric locomotives – an LNER design – head a Merry-go-Round coal train at Dinting on the former GC route through the Woodhead tunnel. (Barry J. Nicolle); *Below* Merry-go-Round empties on the GN main line at Doncaster behind a Class 47 locomotive in 1975. (Rev. Graham B. Wise)

chain of command. Line Management started in September 1957 and ended in July 1964. The measures which it had taken to achieve economies such as the closure of unremunerative lines – notably the Midland & Great Northern Joint Line, 174 miles in length on 28 February 1959 – and the re-organisation of freight marshalling yards and train working had reduced the amount of supervision necessary at all levels so that, sad as it was to see a most successful Line organisation broken up, it had become quite inevitable. The weekend of 3/4/5 July 1964 saw another great physical and mental upheaval and re-settling of staff at Liverpool Street as part of the General Manager's Office from the following day, 6 July.

When Prime Minister Harold Macmillan called on Sir Ivan Stedeford to chair the 1960 Advisory Committee on the British Transport Commission the Group's report led to the break-up of the BTC and the separation of British Railways and the appointment of Dr Richard Beeching as its chairman. But it did not need Beeching to tell the Eastern what to do; he had been forestalled, for good husbandry had already taken action and anything further could be clearly damaging, as indeed it was nationwide. Retrenchment and retraction before but mainly under the Beeching policy reduced the amount of originating traffic from branch lines into the main streams to such an extent that whereas the all-line originating freight tonnage in 1954 was almost identical with that in the pre-war year of 1937 – 283m tons – by 1962 it had been reduced to 228m tons and this was reflected in the ER's through-through or staging services, that is, those freight trains which start up in one Region and pass through the Eastern to terminate in another Region. It is worthy of note that while all Regions lost 55m tons in those nine years the Eastern Region over the same period lost only 2m tons, maintaining a steady average of 44m tons. By 1980 the all-line originating tonnage had fallen to 153m metric tonnes, equivalent to 149m tons.

In the 1964 re-organisation of the General Manager's headquarters the author was appointed Freight Assistant, covering the newly-defined Eastern Region, that is the LT&S, the GE, the GN, and the GC, or that part of it not ceded to the London Midland Region, plus the Midland Chesterfield-Darfield line the ER took from the LM. The GN lines in the

West Riding, in fact the whole of the ER's one-time Leeds District, was lost to the North Eastern Region. So, broadly, the new Eastern Region covered the area from and including the Thames dockland, Fenchurch Street, Liverpool Street, and King's Cross, into East Anglia, up to the Humber and across to the east portal of Woodhead tunnel at Dunford and down to a boundary at Horns Bridge just south of Chesterfield Midland and thus included the north east/south west traffic, the importance of which seldom seemed to be recognised or acknowledged but which was greater than that on any main line to and from London, if suburban commuter passenger trains are excluded. Summarised, the Eastern lost its penetrating lines in the Manchester/Liverpool area west of Dunford, the whole of the Nottingham district, penetrating north west of Aslockton, and the whole of the Leeds district north of Doncaster. The gain was the Rotherham district from the London Midland.

Traditionally, over the many years right up to the end of Line Management in 1964 the Head of the Freight Section or the Freight Assistant as the case may be, was responsible for freight arrangements, train working, transits, and any forward planning in connection therewith, including the freight policy side of train timing, locomotive diagramming, and the locomotive building programme for the years ahead. Any new services to meet traders' requirements, or alterations to existing services, were arranged with due cognisance taken of locomotive and crewing costs. The desirability of re-timing a freight train, if possible, to secure engine and/or crewing benefit was referred to the Freight Section by the Timing and Diagramming clerks for adjudication because of the overall effect on service and transits; the best possible transit for customers' traffic was, after all, the raison d'être. The new 1964 re-organisation changed that by making the Central Timing & Diagramming office the planners for train working, leaving the Freight Movement to monitor the service. This was a great mistake. Experience has shown that the best train operation came from,and was the result of, the same people being responsible for both planning the trains and seeing that they run to plan for therein is a continuing personal involvement and not divided responsibility. The man who plans and executes becomes a dedicated craftsman; the man who plans and leaves the rest for someone else to monitor detaches himself from wholehearted interest. The evidence for this is in the statistical and service performance and, in no less degree, the impairment of relationships man to man. True, it might be pointed out by a supporter of the 1964 re-organisation that the unification and co-ordination between planning the service on the one hand and operating the service on the other hand was through the Movements Manager, but it did not work out like that. It is important to stress the craftsmanship inherent in conception, planning, implementation, monitoring, and post mortem if necessary, by a Freight Officer for freight and by a Passenger Officer for passenger traffic. Had this been done in the implementation of that post-Beeching project, the National Freight Train Plan of 1967, which was imposed on the Regions, there would have been less disruption and a great deal of traffic would have been retained by rail.

9
MERRY-GO-ROUND

Close association between the Eastern Region and the National Coal Board coincided almost automatically with a similar attachment to the then Central Electricity Authority, later to become the Central Electricity Generating Board, and to a lesser degree with the then British Iron & Steel Federation, mainly in connection with the supply of steel-carrying wagons to the various works at Scunthorpe. The electricity industry, like the Coal Board, and like the railways, in the 1950s were beginning to rcover from the ravages of war, to spread wings and develop for the future. Town gas was becoming unpopular and, redolent of the gas-jet mantle era, old-fashioned, and was being ousted by the flick of an electric switch. Some of the smaller gas-producing plants sponsored historically by town councils were becoming life-expired and aromatically abnoxious. The new Gas Board carried out a rationalisation programme and with the new means of production from oil and North Sea gas becoming available, it was less dependent on coal and re-appeared on the energy scene with 'modern gas'. Gas and electricity became competitors for domestic central heating, a feature becoming standard in new house building and conversions in the 1960s.

Electricity consumption boomed and the reason was not difficult to see. On the domestic side, particularly where both husband and wife went out to work it was a convenient and clean form of lighting and heating. The gas mantles, matches, sticks, firelighters, coal and resultant ashes still common in the 1920s and 1930s were on their way out by the 1950s; merely the flick of a switch for both light and heat at 186,000 miles per second! Refrigerators, cleaners, washing machines, irons, television, and a host of other household aids came on the market and caught on in the New Society. Neon signs and advertisements appeared generally. The farmer, tired of his storm lantern, and within the safety of subsidies, could afford to be connected to the power lines which rapidly straddled the countryside. This growth in demand, all within the space of a few years, inevitably caught out the mainly corporation-owned small electric undertakings existing before nationalisation. The result of continued high running and reduced level of maintenance of power plant in wartime Britain, exacerbated by an inadequate number of mainly pre-war power stations to meet demand presented by the severe winter of 1947, placed the electricity suppliers under great strain which led to load shedding – a reduction in the voltage or frequency of supply. Industry had become an electricity consumer on a large scale. Factories, including some of the great steel works at Sheffield, switched over from coal to electricity as their primary energy source. Light industries from the recognised Northern and Midland industrial areas moved south and added to the load growth of electricity. More railways were being electrified. The electricity supply and manufacturing industries generally became major growth industries. The small municipal power stations inherited by the CEA became outdated, inefficient and uneconomic; a grand plan was thus outlined by the proposed siting of large coal-burning power stations on or near the major coalfields and adjacent to rivers having sufficient flow to meet their cooling water needs. The main areas for the concentration of these power stations were along the River Trent in Nottinghamshire and in Yorkshire's Aire Valley.

Initial contact with the CEA was in the late 1940s with the planning of Keadby power station on the west bank of the River Trent near Scunthorpe. The discussions centred on the kind and size of rail layout desirable for catering for the expected level of tonnage. This was in the days of the steam locomotive and, in order to obviate engine crews having to return running tender first, which was an uncomfortable experience for enginemen as they met the elements with coal dust blown from the top of the tender, the CEA was asked to put in a triangle so that locomotives could turn, and they did! Such a facility had never been provided elsewhere on private property but later the North Eastern Region at York got a triangle laid in at Consett Iron Works to ease the working of the Tyne Dock-Consett bulk ore trains. Keadby power station started operation in April 1952 but the triangle for engine turning became redundant when diesel locomotives were introduced years later.

The setting up of the national electricity transmission grid and the creation of the Central Electricity Generating Board resulted in the acceptance that for over 45 miles it was cheaper to transmit electrical energy in bulk through the overhead lines from new power stations in the main coalfields to consumer areas on the grid. Over shorter distances it was cheaper to move the coal by rail from colliery to generating station and to this end plans were afoot for a series of new stations in the Trent Valley. The author was asked for views on the likely sites which the CEGB had in mind, not forgetting the desirable proximity of coal and water supplies. A site suggested immediately east of Newark (Northgate) station in the angle formed by the East Coast main line where it is crossed on the same level by the Nottingham-Lincoln line was quickly turned down, as being unsuitable and was countered by a suggestion for adopting a site near Fledborough station on the Chesterfield-Lincoln line, known to most as the 'LD Line' or the 'Dukeries Line' (the former Lancashire, Derbyshire & East Coast line, later part of the GC). In 1954 the railway representative an CEGB planners agreed the site which became known as High Marnham. The railway layout agreed gave a facing reception line to provide prompt clearance of main running lines, and an internal engine line to enable locomotives to run round trains. Having propelled the loaded train into one of the 14 exchange sidings, each having a capacity of 68 wagons capable of holding a total of 952 loaded or empty wagons, the locomotive would leave with a train of empties. The station was completed and started consuming coal in 1959 but was not opened officially until 5 October 1962. High Marnham became Britain's first 1,000mw power station and the first coal-fired station of such capacity in Europe; 9,000 tons of coal, estimated to keep 5,000 miners at work, would be consumed each day and arrive on 12 trains each comprising 34 24½ ton hopper wagons. The CEGB's method of handling railborne coal was for the CEGB's site locomotive to divide a train and haul it in small rafts of wagons over a below-track-level hopper where staff manually levered the bottom doors open to release the coal and close them again before departure. This was a gruelling task, especially in wintry conditions when the hopper house was like a wind tunnel, and its impact on the author considerably influenced the suggestion for automatic operation at a later date when West Burton power station was considered.

From time to time in post-war years vague reference had been made to the possibility of a power station being erected near Snaith, north of Doncaster, south of Selby, and west of Goole, but the first positive approach from the CEA came when it called a meeting at York on 27 August 1957. From these preliminary discussions it transpired that the site in mind would be in BR's Eastern Region and not in the North Eastern Region as was at first thought. Consequently, the next meeting was at Doncaster power station (not rail-connected) a few days later on 4 September. Before going to that meeting it was suggested to the author by G. F. Fiennes, Line Traffic Manager (Great Northern), that 'the ideal railway permanent-way layout would be a circle or oval'.

Such a design would take the place of the more conventional group of sidings providing for a stock of loaded wagons and, likewise, empty wagons awaiting clearance back to collieries. Fundamentally, the idea was the same as that in operation for the imported iron ore traffic from Tyne Dock to Consett Iron Works from 1953, but much more complex. The Consett working was a movement from A to B and back again in a shuttle, but the number of supplying collieries to furnish a tonnage of five million or more tons a year could easily be 20 or more, with variations in quantity and period, particularly during colliery holidays; over a 52-week year the NCB produced coal in about 48½ weeks. Historically, miners had, as in other industries, worked a six-day week but for some time production had been reduced to five days, while the railway transported in six days, and the power stations traditionally picked up coal from stock, or emptied loaded wagons standing at power stations, over seven days.

In suggesting the circle or oval layout with permanently-coupled trains as units of operation, which virtually meant only one train on the power station site at one time, the railway made it quite clear that its value was restricted unless the existing pipe-line was altered. If it was not altered it would mean the seven day consumption would cause the NCB to build up and hold stocks of loaded coal wagons over the last days before each weekend; the railway attitude was against a facility for storing trains of coal intermediately which in any case would be a waste of wagon user which could not be tolerated. Clearly, the suggested operational revolution necessitated full co-operation between the railway, the National Coal Board and the CEGB. The NCB, however, showed no enthusiasm, and the CEGB at that meeting rejected the idea of permanently-coupled train working on the basis that the then present siding exchange scheme gave them the advantage of having a mobile stock of coal on hand to discharge when required by the power station.

So, the site of the power station having been chosen, the Eastern Region proceeded to agree the point of connection with the BR running lines. It just so happened that the imaginary boundary line between the Eastern and North Eastern Regions of BR ran through the area designated for the point of connection and it was thus decided that the connection based on engineering considerations would be immediately north of the distant signal on the old Hull & Barnsley line, the distant signal being the standard point of demarcation, so it then became a development sponsorship between the CEGB and BR's North Eastern Region at York. That the power station, to be called Thorpe Marsh, was in the immediate environs of Doncaster, that the physical rail connection was likewise but on a dead-end branch line accessible only from the south, that it was envisaged the bulk – 74% – of the coal would be produced in the Eastern Region's Wath area of South Yorkshire, and that a private siding for the contractors' material was required to be laid in on the east side of the site on the Eastern Region's line at Bramwith, cut no ice with the North Eastern team at York. The result was that the CEGB was put in the absurd position of negotiating separately with the two Regional railway headquarters, York and London, over the two rail conncections. It would have been so much easier and sensible to have adjusted the boundary line northwards further away from Doncaster – really only a matter of yards – where it ought to have been in the first place. The Eastern Region, always maintained it was commercially wrong for people and firms in Doncaster's northern suburbs to be doing their business with the District Offices at Leeds and York instead of with the District Office on the doorstep in Doncaster. Equally it was operationally wrong for the Leeds and York Train and Traffic Control offices to control movements on the immediate approaches to Doncaster – up to Shalftholme Junction on the York line and Bentley Crossing on the Leeds line. The York management would not entertain at any time the boundary being redrawn at the three places mentioned, an attitude of obdurate stupidity.

So, for Thorpe Marsh, although the North Eastern team at York pursued the permanently-coupled train idea with the CEGB they negotiated another tandem layout of conventional design like High Marnham with an arrival line, engine run-round, 16 loaded/empty sidings, two brakevan spurs and a cripple siding, to cater for a planned coal consumption of 3.15m tons per year, or 65,000 tons weekly, requiring a fleet of 578 vacuum-braked and 1,157 unbraked 24½ ton hopper wagons, a total of 1,735 wagons at a capital cost at that time of over £3 million.

Following preliminary discussions between the ER and CEGB in 1958–60, plans were revealed early in 1960 for a major development, requiring the consent of the Ministry of Fuel & Power and Nottinghamshire County Council, for the establishment of a new power station on a site covering some 500 acres at West Burton, east of Retford, and close to the River Trent. It would consume some 5¼m tons of coal per annum at full load, commencing with coal going to stock in 1965. Clearly some fresh thinking was called for and the ER team was encouraged when the CEGB said further thought had been given to the idea of permanently-coupled operation, as proposed and rejected for Thorpe Marsh, and that it was proposed to adopt the scheme at West Burton. The CEGB asked 'Would the railway side have any objection?' The Eastern Region was enthusiastic provided that the NCB would co-operate on the following points: first the current five-day production, six-day transit, seven-day consumption, which created a hiatus at one point or another, would have to be broken down by the NCB making block loads, and keeping supplying collieries as few as possible; second the CEGB itself would have to accept coal currently 'as made' by the NCB and pick it up from the station's stock at weekends.

Initially, the intake of coal at West Burton was to be in the order of 1m tons from March 1965 to March 1966, gradually building up to an annual burn of 5.3m tons from 1968 with a possibility of 7½–8m tons annually at some future date. The coal wagon fleet at that time consisted of 16 ton and 24½-ton wagons, some vacuum braked but most unbraked and thus running 'loose-coupled', although there was a transitional stage coupling known as the 'instanter' which would be turned to a short position to bring the buffers close together and reduce the snatches of loose couplings. At the 5.3m tons level it was anticipated some 30 trains daily would be required if steam-hauled, or 20 trains if diesel-hauled. The major developments outlined by the CEGB coincided with BR's changeover of motive power in the early 1960s with a gradual withdrawal of steam and introduction of diesel locomotives, the latter then untried on this sort of work. The main line track connections were to consist of double junctions at both ends of the site with the internal layout in the form of a straight sided looped track or circle, as it was expected trains of coal would approach from east and west, from the Gainsborough and Retford directions.

First thoughts on operation were that trains would go round the loop or circle and pass over a below-track hopper, discharging six wagons at a time on a stop-and-start basis. However, a good deal of physical effort by manual leverage would still be required by the CEGB operators in opening and closing the bottom doors. From previous observations the author was determined that manual door operation would be a thing of the past. At this time the author used Baker Street station on London Transport quite frequently and while waiting for a Circle Line train to approach from Edgware Road became accustomed to watch the changing colour-light signals in advance and, in particular, the automatic action of the trip-cock arm on the track working in conjunction with the signal protecting the junction ahead, the purpose of the trip-arm being to apply the brakes of a train by engaging an arm on the train linked to the brake system should it over-run a danger signal. Seeing the trip-cock rise and fall provided the idea that it ought to be possible to design a wagon which would have its bottom doors opened and closed mechanically by coming into contact with a tripping device between or alongside the rails. The concept was favourably received by the CEGB and later followed up with enthusiasm. Concurrently, and rightly so, it was considered that the 16 and 24½ ton wagons were not big enough and it was agreed to set up a High Capacity Wagon Design Committee led by British Rail, including the author, with the CEGB and NCB thus bringing together engineering and operating personnel.

The BR workshops at Darlington and Ashford developed a wooden mock-up of automatic lineside opening/closing equipment which was tried with an empty 24½-ton hopper wagon at Ashford Works; it was sufficiently successful to justify further tests which were carried out jointly by BR, CEGB and NCB at Richborough power station near Ramsgate on 30 October 1962. There a 'Richborough' wagon, brake fitted 24½-ton No 335606, fitted with normal and disc brakes, had been equipped with the same rotating cam operated door mechanism as used in the first experiments. The test wagon had been thoroughly tried out mechanically at Chislett Colliery four days earlier; it was discharged and reloaded for a second trial and behaved perfectly. It was then part-loaded in the middle with very wet slack after and during heavy rain followed by loading with drier slack. Conditions were therefore not of the best (a good thing) though not as bad as frost/snow accumulations would cause. During four days the load had settled by the time it was put over the hopper for discharge. The trackside equipment contacted the rotating cam on the wagon, the doors opened and the load was efficiently discharged. Passing on to the door-closing equipment the forward and trailing hopper doors closed but the twin middle would not and inspection revealed coagulation of small coal on the door surface rims and this had to be cleaned off manually. The wagon was re-loaded on three occasions providing four test discharges from slack on to an adjacent stockpile. The loads were wet but on each occasion the trip mechanism and doors co-ordinated perfectly. The trackside mechanism provided not only for opening and closing the doors but included safety devices to guard against a wagon rolling back.

The tests were to establish principles of satisfactory mechanical discharge, to establish the final trackside mechanism, in effect the evolution from the originally conceived trip-cock via 'Heath Robinson' gadgetry into a final idea for the drawing-board. It was agreed with the CEGB that although the discharge *time* did not enter the equation at this stage the wagon was timed passing the discharge mechanism (time from buffer to buffer) as 13–15 seconds during which time three sets of doors were opened. On this basis, at West Burton with lineside discharge equipment installed at intervals along, say, three hopper length locations, three wagons could be discharged simultaneously, with coal fairly evenly distributed in the hopper length in 15 seconds. Thus a train could be unloaded in approximately four minutes. In reality more time than this would have to be legislated for to cover contingencies and clearance from the hopper by below-hopper conveyor belts to the power station bunkers, or to stock as required. The representatives were impressed with the Richborough demonstration and its potential and it was agreed the equipment should be fitted to 24½-ton wagons if necessary in the short term but, meanwhile, development of a higher capacity should be progressed quickly. This the Wagon Design Committee set out to do.

The NCB, though, clearly had many problems, and one was the five day week. It was pointed out that the collieries were not set or planned to deal with complete train loads either for loading or weighing and wagons could only be loaded in rafts of up to eight wagons at a time which would necessitate trains being broken up. Furthermore, screens were provided for loading slack and other grades of coal concurrently and therefore pits would not be able to concentrate on loading slack only for the CEGB at any one time. Additionally, some colliery screens had a low vertical clearance, having been erected many years earlier in private ownership days when the height of prevailing wagons was lower. Even the ability to accept 24½-ton wagons was restrictive. Nor was it the normal practice to tie up the whole output of individual collieries to one power station and it was anticipated that the output capacity of the pits generally to meet the West Burton demand of 5.3m tons annually would necessitate coal being drawn from 30 or more pits. It was abundantly clear that the NCB had genuine problems but was stressing them more because it could not readily see what adventages and benefits any changes would bring to the Coal Board.

The CEGB's development programme hit the railways in the days of the steam locomotive working so that initial thinking about methods of operation was tied to that 'horse', and to the existing wagon fleet; BR had a variety of surplus wagons, mostly unfitted stock, that is, without vacuum braking. It was under these conditions that the Eastern Region carried out trials to test the behaviour of locomotives, fitted and unfitted wagons, the over-run and the roll-back of wagons on varying gradients. The latter was a crucial requirement if discharge was to be through bottom doors into an underground hopper, in rafts of six or eight wagons at a time, stopping and starting. The advent of the diesel locomotive brought further problems and the need for renewed trials, using a 2,750hp unit hauling 42 24½-ton wagons established that we could move 15 trains Monday, 21 trains daily Tuesday to Friday, and 13 on Saturday, a total of 112 trains weekly.

The author's suggestion that the locomotive should haul the train over the hopper at slow speed with the wagon doors being opened and closed mechanically raised the question 'what was a slow speed?' . . . 'how slow?' When the 2,750hp Brush Type 4s (now the Class 47) became available it was found that while it could have an almost nil speed from a start, or to stop, it could not sustain a speed lower than 2½mph and that would given an unacceptably high discharge rate causing overfilling due to the inability of the conveyor belts to clear the coal from the hopper. Brush Traction Ltd, builder of the locomotive was asked to design and fit auxiliary gearing on a locomotive for testing purposes. Innumerable train trials were conducted on branch lines in the Doncaster and Sheffield districts to test the hauling and braking ability of an entirely new kind of locomotive. The behaviour of vacuum-braked and unbraked wagons on rising and falling gradients of varying severity at the slowest possible speeds was also tested and it duly emerged that it was desirable, when approaching the 178ft (six wagon lengths) hopper, to have a rising gradient of about 1 in 260 and to travel over the hopper at ½mph; under those conditions the wagons would remain tight-coupled and not buffer up – a feature to prove of significant benefit when fully automatic weighing was introduced. One great difficulty in assessing locomotive capability and behaviour was the gradually diminishing gross load as the train discharged at the slow speed over the hopper.

This was an exciting early-1960 period of development – a period considered to be of significant railway importance heralding the greatest breakthrough in bulk handling on rail. The other rail freight development of the period, the Freightliner, is but an extension of the permanently-coupled coal train principle – special type wagons with specially adapated terminals to facilitate turnround.

The next move was to convince the NCB of the advantages of train-load operation with automatic discharge. But what about automatic loading at collieries? The CEGB had been supplying BR with its longer term coal burning projections for all the main new 200mw power stations, accounting for some 25m tons a year. With a contraction of domestic consumption and rationalisation of coal producing pits, this figure could not possibly represent new tonnage but was largely a switch from domestic/industrial fuel to 'fines' or slack for electricity generation. No one colliery, or big group of collieries could, it seemed, produce fines/slack in that quantity – slack normally being coal that has become degraded by handling through the screens. With 5, 10, 15 or 20 million tons of coal going to new power stations in the Trent Valley alone it was obvious that some collieries would 'make' only one quality and send it all to a power station, therefore, why not do a West Burton in reverse? . . . Why not have an overhead bunker at the colliery from which trains could be loaded and in about the same time that it would be emptied?

Normally the only storage of coal after leaving the colliery was in wagons, railway wagons, and a wagon stock at every colliery, empty and loaded, represented a colossal lock up of railway capital. The only way to achieve real success with the coupled concept at the colliery end was for the NCB to provide a stockpile in an overhead bunker of some 2,000 to 3,000 tons capacity with some financial concession from BR. Obviously this was most advantageous where a particular colliery was able to channel all its output to one particular power station.

It took time though for the idea to get support from the NCB. But in November 1961 the NCB's Deputy Production Director of the East Midlands Division wrote to the Eastern Region:

> At a small number of collieries in the East Midlands Division it is suggested for the future that all output shall be supplied to the CEGB in the form of small coal and this had led to a study of the possible simplification of traffic arrangements on the colliery surface. The general approach has been to consider providing suitable stocking facilities in large bunkers with loading arrangements for dealing with coupled trains on a simple track layout with greatly reduced wagon requirements . . .

A meeting on 29 November 1961 between the NCB, the CEGB and representatives from the Eastern and London Midland Regions of BR was not a 'talk-in' as envisaged, but assumed the importance of a full-scale dress meeting. The LMR representatives were completely in the dark, had no briefing, and were quite unprepared – listeners, and almost 'lost'. It was no impromptu affair for the NCB staff had even prepared and tabled plans for changing surface layouts at four collieries, Coton Park, Kirkby, and Bentinck on the London Midland Region, and Bevercotes on the Eastern Region.

Gerry Fiennes, by then Chief Operating Officer at BTC Headquarters at Marylebone Road had been kept informed of progress. Fiennes pushed the proposal at Marylebone Road/Hobart House (NCB) level and in a talk he christened it the 'Merry-go-Round' which it is not, but as such it was written into the 1963 Beeching Report *Re-shaping of British Railways*. Reading Fiennes' book *I tried to run a Railway* one might get the impression the whole idea of power station hoppers and colliery bunkers was thought up and sponsored by BTC HQ at Marylebone Road but, in fact, both aspects were initiated and progressed through many train trials unbeknown to BTC HQ by the Eastern Region. Moreover the author first raised with the CEGB the desirability and practicability of converting High Marnham and Thorpe Marsh, although new station layouts, from conventional to permanently-coupled block train working, and did, in fact, do it with them. That was how bunker loading, or rapid loading as some call it, at collieries, started.

At Anserieulles Power Station near Lille in France, trains comprising 30 wagons with a gross train weight of 1,200 tons, 885 tons net, were by 1963 being discharged automatically by the Arbel arrangement where, after release of a safety catch, the weight of the falling coal, 30 tons per wagon, in three or four seconds, forced open the bottom doors and, at the same time stored energy in a nitrogen hydraulic cylinder to close them effectively again. This was what appeared to be a permanently-coupled system so the tripartite BR-CEGB-NCB High Capacity Wagon Design Committee visited Anserieulles on 30 April. The commitee found, however, that it was not so, as the SNCF locomotive was detached from the train on arrival, the train then being broken up into rafts for discharging over the hopper, hauled by the power station's own site locomotive and at a time convenient to the power station. The wagon doors were, nevertheless, activated by a photo-cell process, and were self-closing, deriving energy from the discharging coal.

Historically, there had been width, height and track limitations at many British collieries, designed years earlier for wagons generally of the 13-ton capacity, then 16-ton and, from 1954, a 24½ ton wagon in selected workings; this latter wagon was based on the then maximum permissible axle load of 17½ tons and, at that time, only nine per cent of collieries could accommodate it.

Following a series of exhaustive tests and close observation of existing practice and discharge problems at High Marnham by those concerned with the permanently-coupled project, the High Capacity Wagon Design Committee, with the assistance of the railway workshops at Ashford and Darlington, eventually came up with a completely new hopper wagon design, taking full advantage of the railway loading gauge, and an increase in the permissible axle load from 17½ to 23 tons. Having regard to payload/tare ratio and the desire to have instantaneous discharge through bottom doors it was found that a two-axle design was superior to the use of bogies. Thus was developed a 45-ton wagon giving a 26-ton load, or 32-ton load if a 'canopy' was used to increase the height, the former being used until the colliery screen height clearances had been improved. The new wagon had superior discharge characteristics such as galvanised interiors to guard against corrosion and slack adhering to it, rounded corners, improved valley angles, six bottom doors with improved hinges taking up the whole of the wagon floor, and with no inside projections whatever. The wagons were equipped with disc brakes activated by compressed air, strengthened 'instanter' couplings, and hydraulic buffers.

One of the problems experienced by the CEGB (and indeed many other large consumers of similar coal) under conventional operation was during severe weather conditions when wagons of slack standing at collieries and in power station sidings had become consolidated or partially congealed by frost. This worried the CEGB which advocated a form of wagon shaker or heat treatment but it was successfully argued neither would be necessary since if the permanently-coupled scheme was to work properly there would be no standage at either colliery or power station and no frost risk.

The suggested innovation of achieving automatic discharge of coal while the train was moving was continuing to be

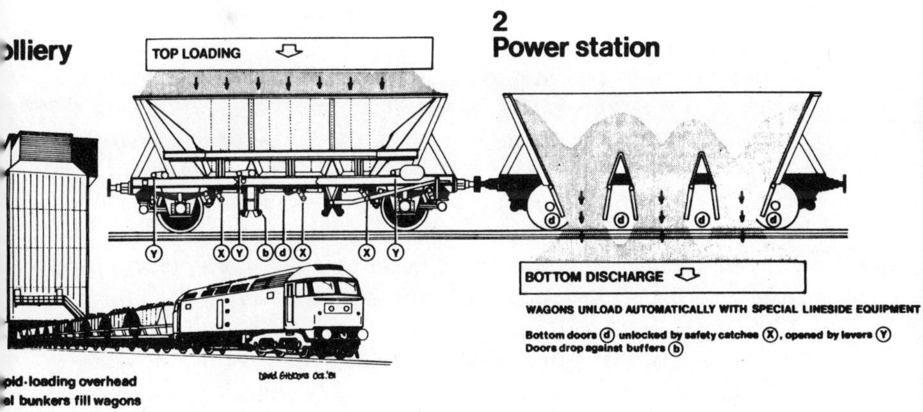

Principle of rapid loading from bunkers while trains are on the move at collieries, and unloading at power stations to form the Merry-go-Round method of operation.

worked and developed by Ashford and Darlington Workshops in conjunction with the CEGB, lineside tripping apparatus engaging the wagons as selected by the CEGB operator using remote push-button control over the length of the 178ft long and two-track width hopper which was to have a capacity of 1,200 tons. The wagon bottom doors were automatically opened and automatically closed. Likewise, equipment was designed and installed ahead and beyond the ground hoppers to release and restore safety catches to prevent accidental or deliberate release of coal on the main line. Obviously, the Wagon Design Committee had to work in unison with the design requirements being worked out jointly by the CEGB and BR's Eastern Region (GN Line) for the operation of the prototype, West Burton. The speed of discharge over the hopper – ½mph – was governed by the rate of clearance on the conveyor belts below the hopper. Meanwhile Brush Traction Ltd had successfully fitted a 2,750hp Co-Co diesel-electric locomotive with auxiliary apparatus to enable it to get down to and sustain a speed of ½mph.

An allied problem of this operation was the question of weighing at the same rate. The CEGB and Eastern Region arranged and conducted trials at High Marnham power station by weighing a train of loaded wagons and empty wagons, while each wagon was standing on the weighbridge, and again while each wagon, loaded and empty, was just perceptibly *moving* over the weighbridge, to determine accuracy and whether weighing was affected by a taut coupling or by buffer-rubbing; the difference in gross weight was marginal. It led to the CEGB to install automatic weighbridges, fore and aft of the hopper, to register gross, tare, and print out the net weights of each wagon.

To comply with railway rules and regulations it was desirable to have a railway wagon examiner to verify the fitness of the wagons to run over the main line back to a colliery for further loading, safety being always of paramount importance. What BR did not want after a train had discharged its load and was standing clear of the hoppers, was an examiner walking down one side of the train and back on the other side to check its fitness for travel. To overcome this potential extra delay it was suggested that immediately on the departure side of the hopper at a point beyond where the lineside equipment would restore the safety catches, a low-level cabin should be provided for a BR man who would obtain an upward-angled view of each wagon as it passed him at ½mph, observing that all the bottom doors had in fact been closed automatically, and that all the safety catches painted white had been closed correctly, also that couplings and pipes between wagons were in apple-pie order. The man could do this without leaving his cabin, aided in hours of darkness by an electric spotlight. This was duly agreed.

Traditionally, according to departmental practice two railwaymen would be provided for an establishment of this magnitude dealing with some 20 trains of coal a day – an Operating Department shunter number taker/checker and a Carriage & Wagon Department examiner with some practical skill; but each would have relatively little to do and might even get bored stiff. So the author addressed the Sectional Council representatives at NUR's Unity House in Euston Road, and also the Local Departmental Committee members at Gresley House, Doncaster, securing their agreement to appoint only one man for each shift and who would be the sole railway representative on the ground and who would liaise with the CEGB men in the hopper house as necessary. Further, the LDC agreed because of technical considerations that the one man on each shift should be recruited from the Carriage & Wagon Department and not from the Operating Department. It was easier for a C&W man to do an operator's work than the other way round and because of mixed duties he would be paid a bonus which was agreed there and then. What could have been a difficult hurdle was negotiated without trouble; this was because the representatives at both levels were completely 'taken up' by the novelty of the project and wanted to be associated with it.

A new modern signalbox was built beside the main running lines at West Burton to control the movement into and out of the power station from in both directions and the clearance distances were such as to provide immediate access. A short cripple siding was put down on the departure side should it be necessary in case of emergency to detach a wagon unfit to go out on to the main line.

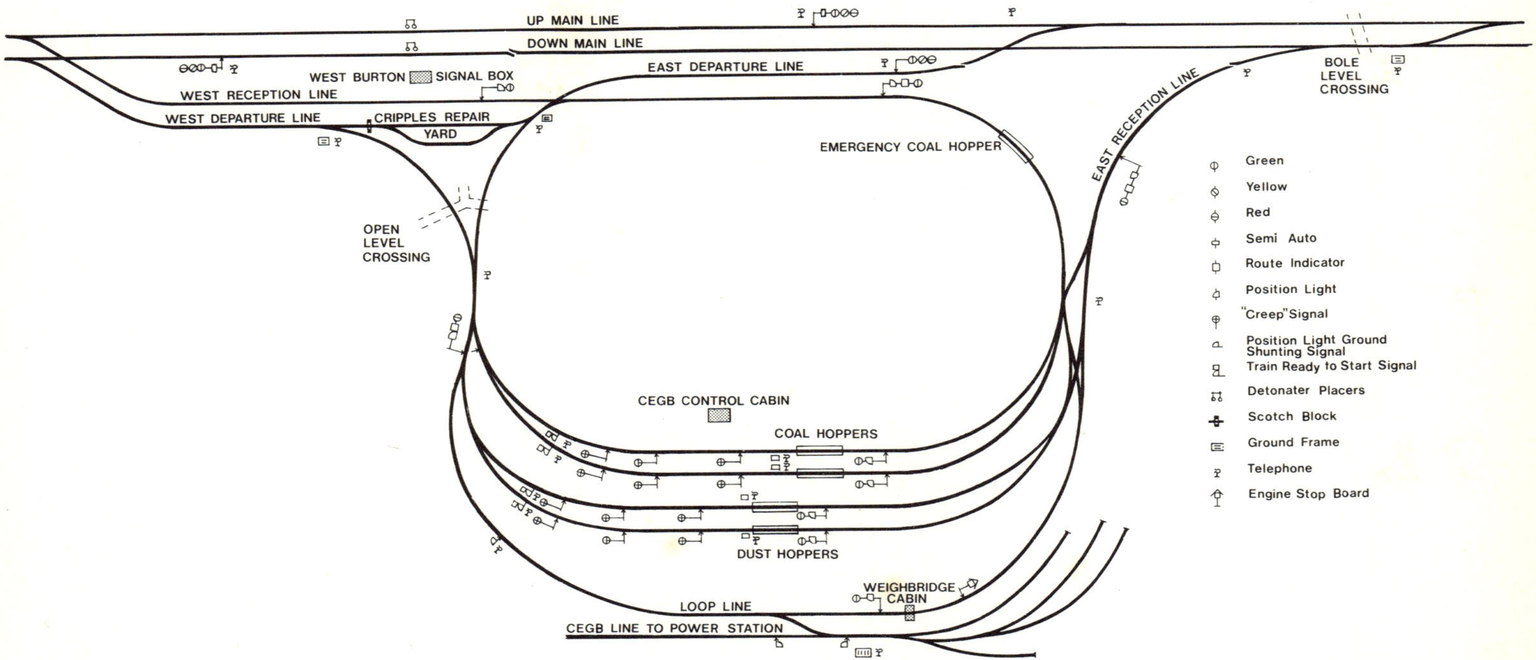

Track layout at West Burton power station for Merry-go-Round operation.

So, from a stop–and–start, stop–and–start, tedious and laborious method of operation was evolved a continuous non–stop working – unless something went wrong. It was, indeed, a truly historic day when in September 1965 the first permanently-coupled or 'Merry-go-Round' train load was discharged in this unique manner at West Burton, then the largest power station in Europe, and an important prototype for coal deliveries. The build-up of coal intake was gradual as the station came on to 'burn' and it called for considerable improvisation, and tolerance, on the part of all three Boards, CEGB, BR, NCB.

During 1966 342 trains were discharged with an average discharge time, including entering and leaving the hopper house, of 44 minutes, and an average total time of the train on the site of 123 minutes which covered various teething and working-in troubles. In one particular week, ending on 10 June 1966, 13 trains with an average payload of 1,013 tons had an average discharge time of only 36 minutes. In the following year, 1967, 1,096 trains were emptied in aproximately the same average time and in the spring of 1968, the intake had touched 100 trains in a week of five days with an average discharge time of 38 minutes, although it has to be admitted that the average payload was below what should have been achieved by the National Coal Board's men at the pit head.

In an article in *The Railway Gazette* of 17 February 1961 the author outlined the first thoughts and agreement reached between British Rail (Eastern Region, GN Line) and the CEGB for a permanently-coupled scheme layout and the research that was still going on to obtain full automation. It provided a great deal of interest in and outside railway circles even to the extent that other railway Regions and some industrial executives were telephoning with enquiries. The valuable information gained by the ER from the many train trials to ascertain the effect over a 178ft long hopper, of buffering-up, over-run, roll-back, of wagons on various gradients, speed, controlling a train with a diminishing load, weighing wagons on the move, frozen coal, discharging sticky and frozen coal, and wagon design characteristics was passed on and the good economic sense inherent in the permanently-coupled project created such enthusiasm that other railway and electricity regions started to copy with their own new power stations being built, in particular Ferrybridge and Cockenzie, to try and be the first to operate the new system. But it was only appropriate that those responsible for the planning and design of West Burton in CEGB and BR Eastern Region having done all the exploratory and proving 'donkey work', should have the honour of setting up the world's first 'Merry-go-Round' (MGR) rail system.

Following these successful pioneering negotiations and satisfactory conclusions, BR with the CEGB, then jointly designed the rail layout for the nearby Cottam power station, the next to be built in the Trent Valley. Other railway Regions collaborated with the CEGB for the new power stations in their territory. West Burton's coaling date set for July 1965 was followed by Ferrybridge 'C', then the others in order:

Eggborough	November 1966
Ratcliffe-on-Soar	August 1967
Aberthaw 'B'	August 1967
Cottam	January 1968
Ironbridge 'B'	January 1968
Fiddlers Ferry	January 1969
Rugeley 'B'	August 1969
Didcot	September 1970

Between coaling dates and 31 March 1971 these power stations had taken in 41m tons by MGR trains. In addition BR and the Scottish Electricity Board adopted MGR at Cockenzie and Longannet, and the conventional siding layouts at High Marnham, Thorpe Marsh, Drakelow 'C', Willington, Staythorpe, and Blyth have been converted thus enabling many miles of track to be recovered and with operational economy. Eleven of the power stations have a rail loop while those converted accept MGR trains on a 'run-round' or 'slip' basis. Now 80% of electricity coal goes by MGR to the major coal-fired stations in trains with a payload of 1,000 and sometimes 1,400 tons as a result of improved loading techniques by the NCB, and with a turn-round at the power station down to one hour.

The project received publicity in many quarters including the Eighth World Energy Conference in Bucharest in July

1971 when a Paper on MGR was given by Dr B. D. Leason of the CEGB Planning Department and who was a member of the team setting up the prototype West Burton.

After the CEGB had accepted the basic MGR concept for West Burton in the early 1960s. MGR operation for Immingham Docks was examined for export/coastal coal shipments and for imported iron ore. Immingham Dock, which opened in 1912, possessed unequalled facilities for coal exports with a capacity of 5,600 tons an hour; to feed the hoists were the largest storage sidings in Europe with a total wagon capacity in the Dock domain of around 25,000. This was taken up in the hey-day of steamships plying round the world. Coal, pit props, grain, timber, wood, iron ore and general merchandise were the staple cargoes, supplemented by seasonal passenger steamers belonging to the Orient Steam Navigation, White Star Line, Canadian Pacific Railway, and others. Dock property covered an area of 1,000 acres with 166 miles of sidings, a sitting-duck for rationalisation. The Portmaster listened to an exposition of permanently-coupled train working, an idea then in its infancy, and a sketch plan was drawn up there and then to illustrate how it could fit into the existing storage-siding area, taking advantage of the several junctions controlled by signalboxes. Export of coal dwindled but the import of iron ore increased to such an extent that a MGR system came into operation in 1974 between Immingham and Scunthorpe with some 13 trains daily each conveying 1,575 tons.

The NCB eventually used the system for coal exports through Immingham and Leith Docks. Other industrial users of the MGR concept are the British Steel Corporation not only at Scunthorpe but also at Ravenscraig and Redcar, Bowaters at Sittingbourne, Rugby Portland Cement at Northfleet, Tunnel Cement at Tring, and Aberthaw Cement at Aberthaw.

Looking at the detail of the colliery end of MGR, Bevercotes was the guinea-pig even though it had been planned for conventional operation. 'Conventional' in this sense means that the layout was designed from practices established over many decades. From the air the layout looks like an hour-glass or egg-timer, with groups of sidings on each side of a bottleneck. Empty wagons are placed by the railway in the reception sidings which are usually on a gradient so that NCB staff can gravitate empty wagons as required down under the screens at the bottleneck and then, when loaded and weighed, gravitate them further into the loaded sidings from which point the BR locomotive clears them in train loads direct to destination or to some near-by marshalling yard. Internal movements within the colliery area other than those by gravitation are carried out by NCB locomotives and men as necessary.

Bevercotes was planned on these lines but when it was opened in 1961 it was with a partial track layout, just sufficient to deal with contractors' material and the early small tonnage of coal, building up to what the NCB hoped would be 1¼m tons a year, and to employ some 2,000 men. Unfortunately, various troubles developed below ground, and hindered or stopped output for a time – a costly business for the NCB and BR because of a nil or reduced return on capital outlay. Nevertheless, the breathing space coincided with the new thinking on layout and operation by doing a West Burton in reverse. The idea had got up to Lord Robens, then chairman of the NCB in Hobart House, and he decreed that as the new Bevercotes colliery was fully automated below ground with the latest mining machinery and belt conveyors with less pick-and-shovel work, it should be automated above ground and was required to be completed in time for the proposed early visit of representatives to the International Mining Congress. Work on an overhead bunker began in earnest. Full automation below and above ground reduced the NCB's anticipated manpower complement from 2,000 to 700 with production up to 1½m tons a year and it would be the World's first entirely remotely controlled coal mine; it was the Coal Board's guinea-pig. Instead of screens spanning a group of sidings a preparation plant, moving belt and overhead bunker was erected. No *group* of empties sidings, or loaded sidings, was required; the rail layout was confined merely to a continuation of the single line from Ollerton, with a run-round loop at Bevercotes, west of the overhead bunker which would feed four wagons at a time in the train positioned by the BR locomotive. Trials were made on appropriate gradients to ascertain the ability of a Type 4 2,750hp locomotive to position wagons accurately under the bunker and were eminently successful. At the power station the locomotive was to discharge a train with a gradually diminishing load, but at Bevercotes the gross weight of the train was increasing by four loaded wagons at a time. Underneath the bunker, with all weighing and discharge movements electrically controlled, it was imperative that wagons remained static under the chutes while loading was in progress and retarders like those in use at Whitemoor marshalling yard were fitted, not to brake the wagons, but to act as wheel grips, coming into operation automatically at the moment of stopping. Loading, weighing, data processing, railway signalling and point operation are electric and automatic.

From the railway viewpoint it was unnecessarily sophisticated, but the word had gone out from Hobart House that it *must* be fully automated. It is worthy of record that anyone seeing Bevercotes from some distance away, or even nearby, could be forgiven if they did not recognise it as a coal mine for it looks more like an ordinary factory with one railway siding. It is clean and smokeless and no blot on the wooded area surrounding it.

Following the establishment of this sophisticated Bevercotes layout some of the local NCB engineers thought about a less elaborate scheme, and eventually produced a prototype bunker at Langwith, a bunker which would load wagons while they were moving slowly underneath – precisely opposite to the West Burton discharge.

It was estimated that British Rail had some £100m capital locked up in wagons being used for storage in collieries and in 1966 an agreement was reached between BR and NCB so that if the latter constructed overhead bunkers of some 2,000/3,000 tons capacity at certain selected collieries BR would make some contribution to the capital cost. By the spring of 1977 no fewer than 27 collieries were using overhead bunker loading systems and it was planned to adapt another 26 collieries or so, at a cost to the NCB of around £8m. Eventually the aim was to have 50 pits in the scheme but in addition other mechanical ground-to-wagon methods were to be used. No business concern, nationalised or not, would spend that amount of money in extending a method of working without an adequate return on capital, the extent of which is not easily gauged. A large colliery conventionally laid out and operated had sidings accommodation for anything up to 2,000 wagons, railway and internal user plus shunting locomotives and manpower. With bunker loading one siding plus locomotive run-round would suffice. Early in the 1980s total capacity of the Board's rapid loading installations was 42m tonnes per year. During the NCB year 1980/81, out of 117.7m tonnes total sales, 82.9m tonnes went to power stations.

At the other end, the CEGB estimated at each power station served on the MGR principle there was a capital saving of some 33 acres of land, about 13 miles of track in the form of exchange sidings and maintenance, and shunting engines throughout the 24 hours seven days a week. Moreover, the NCB's five-day production and the CEGB's seven-day consumption rate inevitably meant that on a conventional scheme most of the weekend's 'burn' was in the form of a buffer stock held in BR wagons – wagons with wheels being used as storage bins – thus locking up some £200m of railway capital. A power station the size of West Burton, and others which have since been built, taking in 5m tonnes a year would, if laid out and operated on conventional lines, require some 2,000 wagons to service it, but on the MGR system only about 200. So BR gained tremendously on wagon user; an early calculation was that a 24½-ton wagon would produce 27,600 net ton miles annually on a four-day turn round, 20-mile loaded basis, whereas the projected 32½-ton wagon for MGR would produce 614,000 net ton miles. Thousands of older, smaller, non vacuum-braked wagons became surplus to requirements and maintenance was saved by their withdrawal. By October 1981 BR's stock of MGR wagons has increased to 10,565 with another 460 in the 1982 Building Programme. They are capable of running at 60mph loaded and empty and are scheduled to replace 21-ton fitted hopper wagons in trunk haul house coal concentration workings. The average wagon load of 7.78 tons in 1947 has more than trebled, such a productivity achievement stemming largely from the introduction of MGR working.

The economies to the three Boards, NCB for rapid loading, BR for speedy transit, CEGB for instant discharge, together with the overall saving in wagon-capital, really defy calculation but the aggregate figure must run into many millions of pounds; it does not end there, for the MGR system has been adopted or adapted for other traffic flows. There was thus an undisputed case that not only should new power stations be designed on the MGR principle but that existing ones should if possible be converted. At High Marnham and Thorpe Marsh the limited land area precluded a circle of oval track layout but a straight road and locomotive run-round was retained, enabling much permanent-way to be recovered and used elsewhere, and dispensing with the CEGB's internal shunting locomotive and men. By 1981 virtually all major power stations had been converted to MGR working.

Although the MGR system is an undoubted success it has to be admitted the system has not worked out entirely as conceived; on average the MGR wagon is getting only about 25% of the journeys a week originally envisaged. Without doubt the terminal time at each end, loading and discharging, is reasonably satisfactory where bunkerage is provided but can be protracted where it is not, and where geological problems cause variations in production and preparation. Mechanical breakdown at power stations upsets the cycle, and railway crewing and locomotive availability problems also cause staging of MGR trains at intermediate points – a feature which is absolutely contrary to the concept. As soon as the NCB can reduce the number of pits supplying any one power station by coal preparation, the true effectiveness of the MGR scheme will be achieved.

Nevertheless, the negotiations between the three Boards at all levels in connection with West Burton, the first of the Merry-go-Round power stations, spanning a period of seven years, resulted in the fostering and cementing of a happy working relationship, with all staff proud to have had a hand in such a venture. What was done as the result of efforts during the five years or so during the conception of the scheme by BR Eastern Region and the pioneering, initiative and enthusiasm exercised by the BR-CEGB-NCB team who did all the thinking, designing, testing, and proving that the idea was attainable and efficient, from basic to sophistication, from the uneconomic to the greatest money-saver that bulk transport has ever known.

Pulverised fuel ash

During the early stages of planning coal deliveries to West Burton power station the CEGB approached the Eastern Region about a decidedly important side-issue, or what could more aptly be called a by-product, not only from the projected West Burton power station but also other new power stations planned for the Trent Valley – the problem of disposing of the ash residue known as pulverised fuel ash or, more commonly, fly-ash. In the older and smaller power stations this had been no great problem, because quantities were so much smaller, but in the 1960s, with a proposed annual coal intake of from 5 to 8 million tons at West Burton alone, the resultant ash for disposal would be in the region of 1 to 1½ million tons annually, or 4,300 tons daily. The combined ash residue from West Burton, Ratcliffe-on-Soar and Drakelow 'C' power stations was expected to be about 10,000 tons a day for which there was no commercial market. True, small quantities could be sold, but not 10,000 tons a day. The CEGB had three possible destinations in mind for dumping the ash, but two involved purchase and construction of several miles of branch line to reach the site and were thus unsuitable. Back in the post-war period in 1949 the author tried to get a daily trainload of rubbish from three London boroughs, Marylebone, Paddington and Westminster, to be dumped in the worked-out claypits of the London Brick Company at Calvert, on the Great Central main line north of Aylesbury. That had not materialised but why not do something similar now? The possibility of filling in the London Brick Company's claypits at Fletton on the southern side of Peterborough was suggested to the CEGB. Another problem was one of speedy loading with regularity and reliability of service at the power station, and an equally speedy discharge at the destination tip, with an effective method of preventing dust nuisance at each end from a commodity not previously carried by rail and whose characteristics were unknown. The ash was a grey powder as light in weight as lady's face powder, and which, when dry, was capable of being lifted by winds and dispersed over a wide area. It had a 'packing' property somewhat similar to that of cement; as a result it was decided to design and develop a new wagon with pressurised discharge. The CEGB was receptive to the Fletton suggestion and the two sides, the CEGB and the London Brick Company, were brought together with the Eastern Region. Outline planning consent was obtained at the end of 1962, and earth moving on a huge scale started in May 1963 for the construction of the necessary embankment and a reservoir within the rail loop.

At the initial survey following the suggestions advanced in 1960, the capacity of the pits was estimated at 30 million tons of fly ash, but because clay excavation was a continuing process, the capacity before filling catches up with excavation may rise to over 50m tons.

As the prototype MGR power station at West Burton was being planned 'in the round' for coal discharge, it was agreed that the ash loading plant likewise would be on a parallel track. Similarly the rail layout at Fletton was to be 'in the round', so that the locomotive on a permanently-coupled ash train, consisting of 48 21-ton 'Presflo' wagons, specially designed for this traffic, does not leave the train except for its own servicing. Entrance to and exit from the unloading site was under the control of the then existing Fletton Junction signalbox. The diagrammed working of a train, diesel locomotive, and men starts from New England yard, running empty to West Burton, loading up, circling the loop, returning to Peterborough and into the site at Fletton, again progressing round the loop for discharge and returning to New England yard for the next run.

Filling the claypits with ash started in 1965 when West Burton power station went 'on burn'. At Fletton the CEGB's own operating staff connect compressed air hoses to the Presflo wagons, the ash is forced out to a slurry mixer, where the addition of water creates a wet slurry, the water being supplied by pumps from the reservoir located within the rail loop. The slurry is discharged into channels from where it is removed by slurry pumps, with a total capacity of about 9,000 gallons per minute, and capable of delivering the slurry a distance of 1½ miles to the more distant pits. After a pit has been filled and allowed to consolidate for some months, the surface is covered with a minimum depth of six inches of top soil, obtained from the nearby sugar-beet factory which has an excess of such soil (obtained from washing the beet) running into thousands of tons a year.

In the early stages of the Fletton project the CEGB engaged the services of a prominent landscape consultant. Tree planting is a continuous process, planting as and when sites became available. As to the ultimate future use of the 3,000 acres of reclaimed land on the periphery of, or within, the boundary of the City of Peterborough, plans envisage a new by-pass road, development for urban building and light industries, playing fields, an amenity lake for sailing etc with the remainder going back to agriculture. As a conservation measure so close to urban life the land value after reclamation must be enormous. During the 30-year plus life of the scheme some 50m tons of ash will have been deposited from three MGR power stations, West Burton, Ratcliffe-on-Soar and Drakelow 'C'. Of the 3,000 acres of excavated pits some 307 acres had been filled in up to October 1981; of the 307 approximately 140 acres at that time had been handed back officially by the CEGB to the London Brick Company, since when crops have been grown and sheep and cattle grazed. According to the London Brick Company's Annual Report:

> 'The scheme is one of the most exciting environmental projects to be undertaken in Britain, for it uses on a grand scale, the waste product of one industry to make good the effects of another and so return what might otherwise be derelict land to new and productive use.'

Introducing the CEGB to the possibilites of dumping at Fletton and bringing the two parties together resulted in the LB Co not only making an agreement with the CEGB in 1963 but it caused the Brick Company to realise it was the first major scheme in which suitable filling material has been available in sufficient quantities to enable large scale reclamation to be achieved. The immensity of the long term operation obviously had a considerable impact and opened the eyes of the Brick Company to the enormous potential that existed for industrial and domestic waste disposal and subsequent land reclamation. Thus a subsidiary company was formed – London Brick Landfill Development – to negotiate with industrial firms and local authorities. This it has done with considerable success, one project being the removal of 1,200 tons of domestic rubbish by train daily from North London to worked-out claypits in Bedfordshire. Far too many years have elapsed since the representatives of the three london Boroughs at Marylebone, Paddington and Westminster, were taken down to meet the London Brick Company's staff at Calvert in July 1949. The three boroughs did not take up the idea of filling the Calvert claypits as meanwhile they had found spare capacity on the Thames barges to Barking Creek. However, the wheel has turned full circle. Following publicity in recent years about the abortive 1949 attempt the idea was revived and the London Midland Region instituted a service carrying rubbish from North London to Calvert 32 years later! It was appreciated in 1949, and many times since, just how valuable a hole can be for a modern developer; an entrepeneur could make a fortune! As for the CEGB and the London Brick Company they have acknowledged the author's contribution at the Brick Company's Annual Meeting in 1981. The railways' part in this? Up to the end of October 1982 16,000 trains have moved more than 15,000,000 tons of fuel ash to Fletton, with much more to come.

10
THE RAPE OF RAIL

The original Nationalisation Act of 1947 envisaged full integration between road and rail but this ideal was never achieved because of clashes between business interests, the road lobby, the railways and the unions. After numerous transport reorganisations road haulage became free of the nationalisation restrictions within a decade or so and 15 years after its formation the British Transport Commission was disbanded. Hiving-off became the antithesis of integration; it was no answer and never will be an answer. It was just another organisational exercise which provided more scope for management proliferation. There will never be a co-ordinated rail/road policy while hived-off interests like Freightliners, the National Freight Corporation, and the National Bus Company exist. With privatisation there is no chance of integration or even co-ordination.

There can be no doubt whatever that the Marples/Beeching get-together of 1962 and Beeching's Re-shaping Plan of 1963 set the seal of twilight on railways as we knew them. Marples' attitude when travelling by train as Minister of Transport was frequently reported as being one of arrogance. He was no friend of the railways; his thoughts, intentions and actions were set on motorways. The motor industry, road and bridge construction, catering contractors, and property developers had a stronger parliamentary lobby than British Railways. A further disadvantage was that in the period from 1947 to date there have been only eight BTC/BR chairmen against 18 individual Ministers of Transport, the latter appointing the former. Since the second world war the nine individual Prime Ministers have never provided continuing governmental consistency and firm policy support for the state-owned railway system.

BTC/BR Chairmen

1947–53	Lord Hurcombe
1953–61	General Sir Brian Robertson
1961–65	Dr Richard Beeching
1965–67	Sir Stanley Raymond
1967–71	Sir Henry Johnson
1971–76	Sir Richard Marsh
1976–83	Sir Peter Parker
1983–	Sir Robert Reid

Prime Ministers

1940	Winston Churchill
1945	Clement Attlee
1951	Winston Churchill
1955	Anthony Eden
1957	Harold Macmillan
1963	Alec Douglas-Home
1964	Harold Wilson
1970	Edward Heath
1974	Harold Wilson
1976	James Callaghan
1979	Margaret Thatcher
1983	Margaret Thatcher

Ministers of Transport

1940	John Reith
1940	J. T. C. Moore-Brabazon
1941	Lord Leathers
1945	Alfred Barnes
1951	J. S. Maclay
1952	A. T. Lennox-Boyd
1954	J. A. Boyd-Carpenter
1956	H. Watkinson
1959	Ernest Marples
1964	Tom Fraser
1966	Barbara Castle
1968	Richard Marsh
1969	Frederic Mulley
1970	John Peyton
1974	Frederic Mulley
1979	Norman Fowler
1981	David Howell
1983	Tom King
1983	Nicholas Ridley

'Close branch lines, important through lines, even main lines', was the cry. 'Let's sterilise the railway tracks and build new roads swathing across virgin countryside'! 'Close depots, stations, marshalling yards and build on the sites – jam for developers'! 'It doesn't matter about lead pollution, conservation of rural areas or the thousands killed and maimed on the roads'!

Closing the branch lines and many stations/depots on through lines (depots open 1954 – 2,258, 1969 – 734, 1982 – 339) reduced the traffic on the main trunk routes and where, once upon a time and not two decades away, freight trains could be seen running on parallel lines or occupying every three or four mile block signalling section. On the 76 mile stretch from Peterborough to London in the 1950s, 55 freight trains were run daily but by 1983 there were fewer than half that number and some of those could be called illegitimate, having been taken off the Lincoln-Whitemoor-Cambridge-Temple Mills route. In those days you could count on about 40 freight trains on the run, half 'up' and half 'down' between Doncaster and King's Cross at any one time, day or night; on a recent visit to York and with my eyes glued to the window, I saw not a single freight train running or standing on the outward or return journeys. Over the Pennines on the Woodhead route, electrified in June 1954 to carry 100 heavy freight trains in each direction, as well as 13 passenger trains over stiff gradients, and using the most economical marshalling yard in the country at Mottram with no shunting engines, just gravity, the daily occupation in 1983 was nil, for the line was closed in the summer of 1981. This fall-back in freight movement is the same over all the other trunk routes countrywide with seldom a freight train in sight. Even with higher capacity wagons meaning fewer trains to carry a given amount of traffic, overnight freight hauls by locomotives running passenger trains by day, a large volume of freight has been lost to road hauliers. Blame the Government, blame the Transport Minister of the day, blame Dr Beeching and his henchmen – and blame all those others who didn't say 'boo to a goose'.

The Eastern Region had built up a freight service better than any other Region, with a high proportion of overnight and one-day transits even to such commodities as iron ore. 'Net Ton Miles per Engine Hour' and other statistics were better than those of any other Region and, indeed, better than the 'all-line' figures. The Operating Ratio was the lowest. Traders were given satisfaction, especially under Traffic Management, which provided a combined Operating and Commercial outlook and a 'feeling' for what was required and what was possible in planning, execution and monitoring. But as recorded earlier, the 1964 Eastern Region reorganisation split the train planning from movement operations at Functional Assistant level. It did not work. In the first place the new train planners had little or no background of freight knowledge and experience, no contact or 'feeling' for commercial needs and, in their execution of Beeching's National Freight Train Plan of 1967 would not be thwarted. Representations to those NFTP planners that such and such an idea, scheme or plan, would completely ruin the transit of so-and-so's traffic and which had been won to rail or retained to rail and given satisfaction for years was met with a stony unwillingness to meet any situation; it was always 'The Board says so . . .' and 'The Board says traders will have to accept a two-day transit' completely overlooking that this often doubled the number of wagons in use. G. F. Fiennes, in his book *I tried to run a Railway* said he found on returning to the Eastern Region as General Manager '. . . it was a misfortune that at Liverpool Street the planning people were thoroughly disciplined by BRB policy . . .' It was indeed a pity someone, or two, or three, or more, had not defied the 'good and great doctor'.

Take this one example of many. Over the years the Eastern Region had given great satisfaction to a consignor at Newark-on-Trent and to the consignee at Welwyn Garden City for the nightly transit of some half-dozen wagons of gypsum. Instead of going through New England marshalling yard at Peterborough in the usual way it was arranged for a through express freight from Hull with a little spare capacity to pick up the wagons at Newark in the late evening and to detach them at Welwyn Garden City, 100 miles away, in the small hours of the morning so that they were always there ready for the firm to start work at 8am. In spite of all the protests what did the new 'train planners' do? They sent the wagons northwards from Newark to Doncaster, 34 miles, transferred them from the Down Yard to the Up Yard, and attached them to a through train to Temple Mills in East London which was re-routed off the Doncaster-Lincoln-Cambridge line on to the GN main line and passed Welwyn Garden City on the way. On arrival at the large Temple Mills marshalling yard the train was shunted and the Welwyn Garden City wagons returned on a trip during mid-morning, hours after they were required, having travelled some 215 miles instead of 100! What can a trader do? Double up the amount of traffic in the pipe-line so that there are always wagons on hand for discharge, taking extra capital to do so, or transfer the traffic from rail to road?

The new National planners at the BRB closed the East Lincolnshire line between Grimsby and Firsby, the junction for Skegness, a distance of some 32 miles, and as a result diverted three through trains daily from Scunthorpe to Whitemoor East Anglia off the East Lincolnshire line and sent them westwards via the Rotherham branch and put them through Tinsley marshalling yard at Sheffield. The balancing trains mainly steel carrying wagons from Whitemoor to Scunthorpe were similarly diverted. Each wagon, each train, each locomotive travelled some 70 miles further and it trebled transit time. Was it costed? Even if it had been, what value would be placed on the calculation?

Costing? In earlier days any freight train scheme was always accompanied by a balance sheet showing the net savings in terms of engines, men and mileage, but with the introduction of the Work Study, Organisation & Methods Study, and Costing, instructions came out from the top that any such schemes in mind should henceforth be 'professionally' costed – costed not by an operator, but by a commercially-trained back-roomer who only had the file of papers to work on. On the first occasion a file was handed over the operator was blandly asked 'which way do you want it to go'? implying that it could be so arranged!

How can railway transits/movement be costed without being arbitrary? The fashionable jargon and spurious doctrines of unintelligent guesstimating have been responsible for much freight traffic being lost to road, as well as the dozen heavily loaded fish trains from the Humber ports, which covered long distances, all of which were thrown overboard by BR HQ to the dismay, disbelief and protests of Regional representatives; moreover it was done at a time when the BRB itself was preaching the gospel of full train unit rather than wagon-unit!

The abandonment of caring for wagon-load traffic, the sole concern being train load traffic, and Sir Reginald Wilson's (a former LNER General Manager) one-time dictum 'the way to shunt is not to shunt' is all very well but it has castrated the railways. The Eastern Region's firm policy in freight operation was always to run as many through trains from source as far as possible and it was only the residue that passed through a marshalling yard, large or small. How often, and where, can you get traffic from one trader/utility to another trader/utility in full train loads, on a railway system now denuded of intermediate traffic depots and with few private sidings, to travel over a railway which is rapidly becoming a north-south, or a to-and-from London axis? Where are the important cross-country lines now, apart from the North East/South West Newcastle, York, Sheffield, Derby, Birmingham, Bristol, South Wales, West of England, and South Coast route?

What if the Post Office adopted the same course and closed down sorting offices in rural areas and dealt only in big lots from large customers? Governmental, parliamentary, and civil service bias to road, not making the best use of years of capital expenditure and expertise, and a gross failure to co-ordinate has, in fact, ruined the railways and the environment, to say nothing of the country's balance of payments. The railways own marketing technique has, in the author's opinion, been found wanting; they should have copied the supermarket policy by dealing with as many commodities as possible, rationalise the stocking (shunting and sorting) have a quick turn-over, and provide a speedy service (transit).

Have matters gone too far for politicians to retrace their steps in the field of freight transport? Very largely, yes, because the railway facilities in the way of marshalling yards, goods depots, and so many firms' private sidings have gone with the Beeching branch line closures. There was a time when firms paid the railway companies to put in siding connections and even met the whole or partial cost of annual maintenance. Now, there is a subsidy or grant to encourage a firm to put in a private siding! National Carriers, a key subsidiary of the National Freight Corporation, was separated from British Rail by the 1968 Transport Act and given the role of funnelling essential freight traffic to British Rail; it is incredible that it should have converted rail depots it took over to road traffic, as at Bristol, Sheffield and elsewhere.

Although the Post Office has not reduced its service in the geographical sense as Beeching did for the railway, the quality and frequency of service has been greatly impaired. The railway lost the Cambridgeshire fruit traffic because it had two rates, 'passenger' or 'goods' between major terminals but with no essential difference in the service provided except greater cost. The Post Office by charging first class and second class postage rates, has shed traffic, is handling less, and what is left is costing more per unit to handle or operate and in effect is pricing itself out of the transport market and leaving the way open for other people to make a living out of the distribution of advertisement leaflets and delivering parcels. The one-time virtual monopoly which the railways and the Post Office had between them, either separately or corporately, in conveying the country's merchandise sundries, parcels, letter mail and parcels post, has been whittled down or given away. The Royal Mail or railway delivery vans are now a rarity and that type of traffic has been picked up by numerous parcels services run privately at a profit, in some cases still using the railway for the trunk haul. If the private enterpreneur can do it why cannot the Post Office and the British Rail? They have not learned the supermarket technique. Nationalisation should have led to the most complete form of integration between the respective Boards. For instance, why cannot British Rail and the Post Office amalgamate and operate one parcels service? Why cannot the spare railway ticket windows sell stamps and do other Post Office business? Why not go further and use railway station buildings for post offices as is often done in Europe instead of the latter being on the opposite side of the road? Why is the bus station not part of the railway station as a general rule and not just in one or two more enlightened towns and cities?

Hiving-off, co-ordination and integration have become a mockery. The National Bus Company is a glaring example of going it alone and putting fingers to the nose. The four main line railways used to have a financial interest – £9½m in 1936 – in individual bus companies and with a railway officer on the bus company's Board there was a committed sense of working together to the extent that each advertised the other's services in their respective timetables, and bus services connected with trains. But now? The constituents of the National Bus Company cannot maintain a local service as timetabled and have reduced services to such an extent that many country areas have become more isolated from public transport than they were at the turn of the century, and yet the NBC can extend itself into a National Express service advertising 'cross-country services linking Wales with the Midlands and East Anglia, and the South Coast with the Midlands and the North' in many cases duplicating rail services and running alongside the railway on new motorways.

The implementation of Beeching's policy of closing branch lines was done in most cases only after public protest and inquiries and generally only when an alternative bus service was available, maybe under subvention. But there was a gross betrayal. Personal experience of one case was the closure of the Taunton-Dulverton-Barnstaple line arising from which a bus service was provided which then made a ½ mile detour to call at Taunton railway station approach. The author used this on many occasions to make the 21 mile run to Bampton, having travelled from Paddington, until the bus service was withdrawn. The only public transport left then was the Royal Blue coach, two services a day running between Bournemouth, Yeovil, Taunton and Barnstaple which called at Taunton bus station and not at the railway station; it necessitated booking a seat, if available, at Taunton bus-station office. The risk of not getting a seat was very high so one stopped going to that part of the world.

The following statistics have a story to tell:

		Mileage of Roads	*Mileage of Rail*		
			Route	*Track including sidings*	*including electrified* *
Pre Second World War					
1939			19,273	50,958	968
1956		187,040	18,965	51,079	1,007
1969		206,126	12,098	32,475	1,969
1982		213,077	10,706	25,690	2,332
1956 compared with	PLUS	26,037	–	–	1,325
1982	MINUS	–	8,259	25,389	–

*included in route miles

Numbers of road vehicles taxed

	1948	*1954*	*1969*	*1972*	*1980*	*1982*
Cars	1,960,000	3,099,547	11,227,000	12,716,000	15,073,000	15,687,000
Motor Cycles & Scooters	?	1,138,629	1,127,400	982,000	1,372,000	1,370,000
Goods Vehicles	?	1,032,868	1,563,500	1,645,000	1,678,000	1,616,000

plus a miscellany of other categories.

	Road Accidents		*Passengers in Train accidents BR*	
Year	*Killed*	*Injured*	*Killed*	*Injured*
1946	5062	157,484	35*	657*
1954	5010	233,271	Nil	486
1955	5526	262,396	40	759
1960	6970	340,581	6	745
1965	7952	389,985	2	344
1972	7779	352,013	6	381
1978	6831	342,904	13	176
1980	6010	323,000	Nil	387
1982	5934	328,362	Nil	150

*Average 1946–50

The near eight-fold increase since 1948 in taxed private cars represents increased wealth in the domestic consumer market, which in itself is a main cause of inflation and road casualties, recent figures being over 6,000 deaths and over 300,000 injured each year and, according to the Ministry of Transport road accidents in 1978 cost £1,614m. Why should the motor manufacturing industry exact such a toll? The car market has had its boom and the industry currently produces an article which is difficult to sell in a saturated market, or cannot meet the competition from abroad. To some extent the car is an economic luxury, and is a lethal weapon. Yet the Government condones it all by granting subsidies!

From 1825 to Nationalisation, railways were constructed and maintained on capital invested by those who had the means and freewill to do so, supplemented by profits accrued from the revenue derived from the traffic put on the railways by those who wished to do so; with both the investor and the user, be it consignor or consignee, it was a voluntary arrangement. The building of high class railway hotels and the construction of the finest docks in the country showed no lack of enterprise and capital expenditure.

The expenditure on road construction and maintenance has been quite contrary to that of the railway. While the owner of a driving licence or vehicle licence makes some contribution to the use he makes of the road system, the greatest proportion of the expenditure is extracted through rates and taxes from people many of whom do not have a licence to drive or own. The total expenditure on roads rose from £388.6m in 1965 to £417.9m in 1966 plus around £200m in lighting and car parks. Expenditure on new construction and improvement of trunk roads including motorways in Great Britain during 1981 and 1982 was £582.5m while that spent on new construction and improvement of principal and other roads was £462.5m It could well be argued that far too much money is spent on motorways and principal roads with no accountability, no calculable return on the investment, while nationalised rail, coal, and docks are expected to do just that. Thousands of acres have been neutralised and the landscape scarred without a great deal of protest because every politician, and many other Tom, Dick and Harry, are owners or drivers and have, therefore, a vested interest. The policy of allowing, and even engineering, a run down of the railway system, constructing more and more roads, authorising the maximum laden weight of heavy lorries to be increased from 32½ to 38 tonnes while, at the same time paying subsidies to rail, road and car industries has landed the Government in an unenviable situation.

The runaway figure of taxed cars, approximately 2m in 1948 to nearly 16m in 1982 is indicative of the social-cum-business trend in self-driving brought about by the withdrawal of, firstly, branch line trains and, secondly, buses. This, and the freedom in which coach operators can set up a service is a further threat to the railway main lines outside the commuter belt. For many the car is a convenience, albeit grossly extravagent but something to be endured for the sake of status; it stands in a garage, in a car park, or outside somebody else's house for the greater portion of its life – a condition blandly accepted by the businessman who would be horrified if his goods vehicle did likewise. It is quite illogical! The weekly cost of maintaining a car has become, even to the lower salary/wage earner a 'must' and is undoubtedly responsible for pushing up wage demands; it is astonishing that many spend more on a car than on the home they own or on the subsidised council house they rent!

After the establishment of the Eastern Region General Management command following the disbandment of the Line Traffic Management organisation which lasted from 1957 to 1964, it was fair to assume that it might last, if not for all time, for a long time at least. But it was not to be. 1965, 1966 and another upheaval was on the way – a merger between the Eastern Region and the North Eastern Region with new Regional headquarters at York.

In view of retrenchment on both Regions arising out of good husbandry before the Beeching regime, the Beeching policy, the National Freight Train Plan of 1967, the big reductions in staff, and 'line of route' operating considerations it was a sensible logical proposition though it did not at first appeal to many staff for domestic reasons. At the same time it was a saddening prospect insofar that so much had been built up over the post-war years, the fused operating/commercial principles, the good statistics, the goodwill, the traffic gained and retained by satisfactory overnight express freights most of which had been or was being thrown away under the National Freight Plan. It was not a process of gradual evolution but rather one of purposeful change regardless. Having thrown so much away by spurning wagon-load traffic in favour of the full train-load unit, overlooking simple arithmetic that numbers add up to a total, then seeking to reverse the trend as in the Board's much publicised announcement that in 1981 it was running 60 Speedlink trains throughout the national network giving overnight transits. A year later it publicised a plan 'to expand Railfreight Spedlink services, especially aiming at traffic requiring warehousing and distribution facilities with maximum trunk haulage of bulk loads beween complexes by rail with local distribution of goods by lorry. Each site would have good road and rail access linking with rail offering regular services. Complexes would include modern warehousing and transhipment facilities'.

There is little or nothing new in this concept of operation. In 1960 the Great Northern Line Management part of the Eastern Region (GE and part of GN, GC, and Midland lines) alone originated 322 Class C fully-braked express freight trains and 553 partially-braked trains all giving overnight transits to 'complexes' but which we called 'goods depots', all having excellent facilities for all classes of wagonload or small merchandise, with fixed and mobile craneage, outside ground storage and warehousing to accommodate a variety of goods – wool, sugar, grain, bananas etc – and some goods depots had bonded facilities for whisky, etc. Marylebone had been a prime example, receiving two trains of whisky from Glasgow every night until the depot was destroyed by bombs during the War.

* * *

Porters have become Stationmen, booking/ticket clerks have become Clerical Officers, fewer Stationmasters have been spread around and re-designated Area Managers, District Operating Superintendents are now 'Divisional' and cover a wider area, Accountants are now Finance Officers and the one-time awesome Staff Clerk has become a Personnel Manager. Maybe this gives an intensified feeling of personal pride and comparability in status with, say, the Civil Service, but where is the traffic, the increased traffic, to manage? In 1960, out of his retirement, Sir Michael Barrington-Ward wrote to the author . . . 'I don't understand or appreciate the organisation now . . . every other person seems to be a Traffic Manager or a General Manager, especially on the LM . . .'!

In this context of change, an admirable article 'Integration versus Fragmentation' written by A. A. Harrison, a former Assistant General Manager at Liverpool Street said: 'When I think of what the railway officers of 40 years ago were expected to shoulder, and did, with conspicuous success for a much greater volume of traffic, I shudder at the fragmentation of responsibility and effort today and what must be the appalling expense of it, not to mention its confusion'.

With most of the staff at Liverpool Street Eastern Region HQ having gone to York to take up their new posts in the enlarged Region, and with others still to go, the author decided to say farewell with very mixed feelings but thankful for a full working life spent during the busiest days of the railway era, in a transition period of steam to diesel and electric traction, from semaphore to colour-light signalling, from mechanised signalboxes to power signalling centres, from pen-nibs copying ink and copying presses to typewriters, carbon paper and duplicators, from the Morse single needle telegraph and buzzer to teleprinters and telex, to radio and computers, but regretful at leaving so many good and friendly people behind. People, yes, – perhaps they do matter more than traffic for 'There is nothing better than that a man should enjoy his work for that is his lot'. (*Ecclesiastes*). And the author did.

INDEX